Endemic Species

Edited by Eusebio Cano Carmona,
Carmelo Maria Musarella and Ana Cano Ortiz

Published in London, United Kingdom

IntechOpen

Supporting open minds since 2005

Endemic Species
http://dx.doi.org/10.5772/intechopen.73421
Edited by Eusebio Cano Carmona, Carmelo Maria Musarella and Ana Cano Ortiz

Contributors
Mónica Moraes R., Carla Maldonado, Freddy S. Zenteno-Ruiz, Alberto González-Zamora, Rebeca Pérez-Morales, Carlos García, Héctor R. Contreras, Javiera Oyaneder-Terrazas, Jiang Zhou, Huaiqing Deng, Carmelo Maria Musarella, Valentina Lucia Astrid Laface, Giovanni Spampinato, Ana Cano Ortiz, Ricardo Quinto Canas, Andrés Cruz-Hernández, Abraham Arellano-Perusquía, Bernardo Bañuelos-Hernández, Luz Urquieta-Ramírez, Froylan Albarrán-Tamayo, Carlos Agustín Aguilar-Ruiz, Marco Antonio González-Márquez, Sergio Junquera-Martínez, Jontari Hutagalung, Linawati Hananta, Nikson Sitorus, M. Soleha, Ngambouk Vitalis Pemunta, Ngwa Donald Anye, Eusebio Cano Carmona

First published in London, United Kingdom, 2019 by IntechOpen
IntechOpen is the global imprint of INTECHOPEN LIMITED, registered in England and Wales, registration number: 11086078, 7th floor, 10 Lower Thames Street, London, EC3R 6AF, United Kingdom
Printed in Croatia

British Library Cataloguing-in-Publication Data
A catalogue record for this book is available from the British Library

Additional hard and PDF copies can be obtained from orders@intechopen.com

Endemic Species
Edited by Eusebio Cano Carmona, Carmelo Maria Musarella and Ana Cano Ortiz
p. cm.
Print ISBN 978-1-83968-252-0
Online ISBN 978-1-83962-353-0
eBook (PDF) ISBN 978-1-83962-354-7

For EU product safety concerns:
IN TECH d.o.o., Prolaz Marije Krucifikse Kozulić 3, 51000 Rijeka, Croatia, info@intechopen.com or visit our website at intechopen.com.

Meet the editors

Eusebio Cano Carmona obtained a PhD in Sciences from the University of Granada, Spain. He is Professor of Botany at the University of Jaén, Spain. His focus is flora and vegetation and he has conducted research in Spain, Italy, Portugal, Palestine, the Caribbean islands and Mexico. As a result of these investigations, Dr. Carmona and colleagues have directed 12 doctoral theses and more than 200 publications among articles, books and book chapters. He has participated in national and international congresses with about 250 papers/communications. He has held a number of different academic positions, including Dean of the Faculty of Experimental Sciences at the University of Jaen, Spain and founder and director of the International Seminar on Management and Conservation of Biodiversity, a position he has held for 13 years. He is also a member of the Spanish, Portuguese and Italian societies of Geobotany.

Carmelo Maria Musarella, PhD, is a biologist specializing in plant biology. He has studied and worked at several European universities, including Messina, Catania, Reggio Calabria, Rome (Italy), Valencia, Jaén, Almeria (Spain) and Evora (Portugal). He is currently Adjunct Professor of Plant Biology at the "Mediterranea" University of Reggio Calabria (Italy). His research topics include floristics, vegetation, habitat, biogeography, taxonomy, ethnobotany, endemisms, alien species and biodiversity conservation. Dr. Musarella has authored many research articles published in indexed journals and books. He serves as guest editor and referee for *Plant Biosystems*. He is a member of the permanent scientific committee of the International Seminar on Biodiversity Conservation and Management guested by several European universities. He has participated in several international and national congresses, seminars and workshops where he has made oral and poster presentations.

Ana Cano Ortiz holds a PhD in Botany from the University of Jaén, Spain. She has worked in private enterprise, in university and in secondary education. She is co-director of four doctoral theses. Her research focus is related to botanical bioindicators. Dr. Ortiz has worked in Spain, Italy, Portugal and Central America. She has published more than 100 works in various national and international journals, as well as books and book chapters. She has also presented a great number of papers/communications to national and international congresses.

Contents

Preface

First of all I want to thank the Editorial IntechOpen for having placed their trust in order to act as Academic Editor of the book Endemic Species. I also want to thank the help given to the co-editors of this book, who have acted generously, using part of their time in this project; and of course my deepest thanks to the group of authors who have participated in this book, without which it would have been impossible to edit.

The value of this book is in its content and in the thematic diversity it contains. It is a book that focuses on various topics concerning the concept of endemicity. It presents studies of endemic plant and animal species from different territories of the world, highlighting the specific richness of the study territories and their conservation status. It also opens up to other health-related topics, such as certain endemic diseases. Readers will enjoy reading updated scientific topics related to botany, zoology and medicine.

This book aims to bring the reader up-to-date topics, mainly about endemic plants and animals of the world, and in particular about those places on the planet that act as hot spots due to their high rate of endemism. Endemism is a phenomenon of stenochoria, therefore it has a distribution area more or less restricted to a given territory.

The first author who used the term endemic in a botanical sense was De Candolle, who applied the term to families and genera that grew up in a single country. Endemic taxa are highly attractive to researchers and scientists, since they are very rare plants or animals that occupy a certain area and are frequently in danger of extinction, a fact that has bothered botanists such as Favarger, who commented, "If an endemic species is destroyed, this represents an irreversible loss, equivalent to genocide." Braun-Blanquet indicates that the study and exact interpretation of the endemism of a territory is the supreme criterion for any consideration regarding the origin and age of the plant population. Endemism is therefore a phenomenon that has always attracted the attention of researchers; for this reason the edition of this work is justified.

I want to thank IntechOpen for having placed their trust in me and asking me to act as Academic Editor for this volume. I also want to thank my co-editors, who have acted generously and donated their time to this project. And finally, my deepest thanks go to the group of authors who contributed to this book.

Eusebio Cano Carmona and Ana Cano Ortiz
University of Jaen,
Spain

Carmelo Maria Musarella
"Mediterranea" University of Reggio Calabria,
Italy

Section 1

Vegetal Biology

Chapter 1

Introductory Chapter: Endemism as a Basic Element for the Conservation of Species and Habitats

Eusebio Cano Carmona, Ana Cano Ortiz and Carmelo Maria Musarella

1. Introduction

A number of chapters have been submitted to the book *Endemic Species* by different authors. Many of them are of interest for conservation, as they deal with species located in special situations, which are in most cases under threat from human activity. Other authors have promoted conservation and exploitation simultaneously, which is an idea of key importance, as these two practices are not mutually exclusive. In fact the reverse is true: using certain species in a reasonable or sustainable way may even enhance the state of conservation.

The chapter entitled *Instrumental Methods for Detection of Lipophilic Marine Toxins in Endemic Species from Pacific Austral Fjords*, containing a detailed description on procuring toxins for certain species of dinoflagellates, is important from the social sanitary point of view. The work on endemisms in Bolivia examines the distribution and megadiversity in the country, with particular emphasis on the Bromeliaceae and Orchidaceae families. The works on endemic flora in Mexico are very important, as is the case of *The Endemism of the Vascular Flora of Mexico Present in Comarca Lagunera, an Agricultural Region in the Chihuahuan Desert*, which studies over 300 species for very special territories such as deserts. These are precisely the areas that will be most seriously affected by climate change. Climate change could be mitigated by means of cultivation techniques and the creation of bioclimatic optimisation models; this is why the work on *Mexican Indigenous Species with Agroecological Uses* is particularly interesting, as it examines certain species that were cultivated in the past and are currently being recovered. The excellent study on *Salvia ceratophylloides* Ard. (Lamiaceae), an exclusive endemic species of southern Italy, alerts to the endangered situation of this species and points to the need for special conservation measures while noting that its threatened status is due to unregulated urban development. In the conclusions, the authors propose the establishment of micro-reserves and the possibility of creating a seed bank and local botanical gardens to favour conservation. These botanical gardens should have a phytosociological character, with a prior study of the plant association or associations in which this taxon is included.

Other territories such as Madagascar have over 10,000 endemic species. Australia has its own species, genera and botanical families, as this continent has three major climate types, tropical, desertic and Mediterranean, along with Central America and many other areas on earth. In areas in Central America, the endemic flora is of particular importance for conservation. The origin of this flora varies in different areas of the planet and may consist of taxa with ancient origins (paleoendemics) or young taxa formed by the evolutionary process after becoming isolated (neoendemics). Plate tectonics and the Alpine orogeny that gave rise to seas, oceans and mountain chains are among the causes of the isolation of populations, which has allowed the genesis of new species. However, the greatest impact on the emergence of endemic species occurred with the glaciations in the Quaternary era, as they took refuge in more southerly environments and subsequently become isolated.

The highest rate of endemism can be found in major mountain chains and on islands in the tropical-subtropical belt. The following are areas of particular interest: Spain, Madagascar, Malaysia, South Africa, Australia, Central America, and particularly Mexico and the Caribbean. Of all the Mediterranean countries in the EU, Spain has the highest rate of endemic plants due to its orography and geological and climatic diversity, as it is situated between the Atlantic Ocean and the Mediterranean Sea and favoured by migratory routes from the northern territories in Europe and Africa. However, Spain is among the countries that suffered the greatest extinction of species during the Quaternary glaciations as a result of the configuration of its mountain systems, which should alert us to the fact that the climate change that is already underway could trigger a similar phenomenon.

2. Methodology

The study of endemism is analysed in several areas and related with the habitat, and fieldwork is conducted in several American and European territories (Mexico, Hispaniola, Spain, Portugal, Italy, Palestine etc.), consisting of compiling phytosociological relevés. The results are various works on vegetation that describes a number of plant associations [1–9], all of great interest due to their endemic character. Other results include works on endemism [10] and a series of studies on the flora of the island published in the journal Moscosoa, together with the Flora of Hispaniola by [11]. This present research is a continuation of the aforementioned works, and we corroborate the bioclimatic and biogeographical studies [1, 7, 12, 13].

3. Results and discussion

Species become isolated through different causes, including geological, geographic, climatic and biological barriers, and throughout history many authors have taken this phenomenon as the basis for understanding the situation of Mexico, Panama, Honduras, Guatemala and the Caribbean Islands—all of which can be considered hotspots and are of great interest for conservation. This is essentially due to the fact that the extinction of species during the Quaternary era was less widespread than in the old European continent due to the configuration of the mountain chains, which acted as corridors for plants and animals. A similar phenomenon occurred with the geological barriers in Hispaniola, part of the Greater Antilles in the Antillean arc. Large mountain chains were formed in the Cretaceous era that was subsequently linked by Tertiary and Quaternary sediments. The Cordillera Central range is siliceous in character and has the highest altitude in the whole of the Caribbean region (Pico Duarte: 3175 m); other mountain chains also originated during the Cretaceous era but have a lower altitude and a calcareous character, with patches of serpentines, which contain a high rate of serpenticolous edaphisms [5, 10].

Examples of this are the islands of Cuba and Hispaniola, which despite their limited territorial extension each have over 6000 plant species of which over 33% are endemic. Five sectors and 19 biogeographical districts have been described for the specific case of Hispaniola, and a series of works have been published on its flora and vegetation [6, 14], highlighting the interest of this island due to its high diversity of flora and vegetation. Of the 6000 species described on the island, a little over 33% are endemic, with a predominance of tree, shrub, vine and epiphyte biotypes; endemic plant species number more than 2000. This high diversity of endemic flora includes their own species and even genera. This is the case of Tortuga Island with *Tortuella abietifolia* and *Salvia montecristina*, which is exclusive to the Morro de Montecristi, *Eugenia samanensis* on the Samaná peninsula and *Pereskia quisqueyana* and *Melicoccus jimenezii* on the eastern plain (eastern Caribbean sector). The geological origins of the Caribbean led to the creation of the Antillean arc, which enabled the migratory route from Florida and the Lesser Antilles and favoured the arrival of species to the large mountain ranges. Their subsequent isolation due to the lack of genetic flow between individuals resulted in speciation and the emergence of endemism.

Climate change is causing territorial climate irregularities and an increase in extreme climate statistics. Certain territories in which species have been adapted to the climate for thousands of years are unable to withstand such sweeping changes in this new rapidly changing climate scenario. It is precisely the endemic species with less plasticity that are most at risk.

However, in spite of the influence of climate change on endemic animals and plants, the greatest threat may come from humans, as human pressure on habitats is increasing by the day in spite of the various protection policies, which are more declarations of intent than actual facts. Agricultural and livestock farming areas are maintained almost all over the planet, and there are very few sustainable farming operations. This is for several reasons: first, the need in some countries today and in the past to obtain food and energy, which throughout history has led to mass deforestation, and, second, the current proliferation of forest fires, which are responsible for the loss of biological diversity. It is evident that endemic species form part of some kind of habitat, which points to the need to protect that habitat, as this will also protect the species. It can therefore be said that endemic species are key to the establishment of conservation policies. In places on the planet that are considered hotspots due to their high rate of endemics, we need to know the plant associations in which these species occur. If this information is not available, we believe it is crucial to conduct research to increase the knowledge of these plant communities. Some authors have studied the vegetation in Cuba, as is the case of [15–18], and numerous diverse endemic syntaxa have been included in the phytosociological classes of *Byrsonimo-Pinetea caribaeae*, *Caseario-Pinetea cubensis* and *Phyllantho-Neobracetea valenzuelanae.* In the case of the island of Hispaniola, where we only know of our own recent studies, there is a wide diversity of endemic syntaxa as a result of the high rate of endemic species [19]. In previous years we studied the high-mountain forests of *Pinus occidentalis* and formulated several endemic associations for the island, *Dendropemon phycnophylli-Pinetum occidentalis*, *Coccotrino scopari-Pinetum occidentalis* and *Leptogono buchii-Pinetum occidentalis*, which are included in the class *Byrsonimo-Pinetea caribaeae*. A detailed vegetation study was carried out in arid/semiarid and dry environments [20], and the syntaxa *Harrisia nashii-Prosopidetum juliflorae* and *Neoabbottio paniculatae-Guaiacetum officinalis* were described—among others—and included in the class *Cercidi-Cereetea.* A similar situation occurs elsewhere on Earth such as in the Mediterranean region, whose characteristic climate is dry with cold winters and hot summers, and whose orography and geology allowed the arrival of Arctic species (Sierra Nevada) and African species (southern Iberian Peninsula) during the Quaternary era. The isolation of species occurred as a result of climate change during the Quaternary era and due to the numerous soil niches that have acted as centres of origin (dolomites in the Sierra de las Almijaras, serpentines in the Sierra Bermeja, etc.). All these Iberian territories that are considered of interest due to their flora have endemic plant associations, which have been prioritised by the EU in its Directive 92/43/EC. This excellent regulation enables these places to be conserved, so it is important to extrapolate it to other places on the planet and particularly to hotspots.

It is evident that the immense majority of endemisms are located within a particular habitat and that this habitat contains one or several plant associations. Paleoendemisms are adapted to very specific environments that occupied greater extensions in the past but have since been restricted as a consequence of climate change, whereas due to their plasticity, neoendemisms may occupy less strict ecological niches and, as noted by [21], be prepared to increase their area of distribution. This implies that the habitat in which these species are located can be catalogued as either paleohabitats or neohabitats. Paleohabitats should be considered as a restricted area, which formerly covered a greater territory and in theory date from the Eocene and Oligocene eras. If the current climate trend was evolving towards the climate of those periods, the paleohabitat would expand, and if the reverse was true, it would disappear and so would the endemic species in consequence. Neohabitats have been generated more recently, appearing between the Miocene, Pliocene and Pleistocene eras, and include endemic species with a greater distribution area. The current climate trend could increase or reduce their area of distribution.

4. Conclusions

Habitat conservation measures must be established to safeguard the flora, and particularly the flora threatened by anthropic actions and climate change. Although human actions cannot be eliminated completely, it is possible to implement a system of sustainable management that mitigates habitat deterioration caused by anthropic action and climate change. This management must have two aspects: the first is local management, which integrates humans in the natural environment by raising educational standards and encouraging the public to value and respect the landscape for which it is essential for people to be familiar with these habitats. It is therefore crucial to have knowledge of plant associations, as certain non-endangered syntaxa can be used and exploited by humans, while others must be strictly monitored for their conservation, as is the case of those we have mentioned previously in this work. In second place, it is important to implement the appropriate national and international policies against fire and climate change.

Author details

Eusebio Cano Carmona[1*], Ana Cano Ortiz[1] and Carmelo Maria Musarella[1,2]

1 Departamento Biología Animal, Vegetal y Ecología, Botánica,
Universidad de Jaén, Jaén, Spain

2 Dipartimento di AGRARIA, Università "Mediterranea" di Reggio Calabria,
Reggio Calabria, Italy

*Address all correspondence to: ecano@ujaen.es

References

[1] Cano E, Veloz A, Cano Ortiz A. Contribution to the biogeography of the Hispaniola (Dominican Republic, Haiti). Acta Botanica Gallica. 2010a;**157**(4):581-598

[2] Cano E, Veloz A, Cano Ortiz A. The habitats of *Leptochloopsis virgata* in the Dominican Republic. Acta Botanica Gallica. 2010b;**157**(4):645-658

[3] Cano E, Cano Ortiz A, del Río S, Alatorre J, Veloz A. Bioclimatic map of the Dominican Republic. Plant Sociology. 2012a;**49**(1):81-90. DOI: 10.7338/pls2012491/04

[4] Cano E, Veloz A, Cano Ortiz A, Alatorre J, Otero R. Comparative analysis of the mangrove swamps of the Dominican Republic and those of the state of de Guerrero. Mexico. Plant Biosystems–An International Journal Dealing with all Aspects of Plant Biology. 2012b;**146**(1):112-130. DOI: 10.1080/11263504.2012.704885

[5] Cano E, Cano Ortiz A, del Río S, Veloz A, Esteban Ruiz FJ. A phytosociological survey of some serpentine plant communities in the Dominican Republic. Plant Biosystems–An International Journal Dealing with all Aspects of Plant Biology. 2014;**148**(2):200-212. DOI: 10.1080/11263504.2012.760498

[6] Cano E, Cano Ortiz A, Veloz A. Contribution to the knowledge of the edaphoxerophilous communities of the Samaná peninsula (Dominican Republic). Plant Sociology. 2015;**52**(1):3-8. DOI: 10.7338/PLS2012491/04

[7] Cano Ortiz A, Musarella CM, Cano E. Biogeographical areas of Hispaniola (Dominican Republic, Republic of Haiti). In: Plant Ecology. University Campus, Croatia: Intech Publisher; 2017a. pp. 165-189. DOI: 10.5772/intechopen.69081

[8] Cano Ortiz A, Musarella CM, Piñar Fuentes JC, Bartolomé Esteban C, Quinto Canas R, Pinto Gomes CJ, et al. Advances in knowledge of vegetation of the Spanish Island (Caribe-Cento America). In: Vegetation. University Campus, Croatia: Intech Publisher; 2017b. -83, 99. (ISBN 978-953-51-5790-8). https://www.intechopen.com/booksprocess/action/chapter/125184

[9] Cano Ortiz A, Musarella CM, Piñar Fuentes JC, Pinto Gomes CJ, Cano E. Analysis of the conservation of Central American mangroves using the phytosociological method. In: Mangrove–Fauna and Flora. University Campus, Croatia: Intech Publisher; 2018. pp. 189-206. Available from: https://www.intechopen.com/booksprocess/action/chapter/128266

[10] Cano Ortiz A, Musarella CM, Piñar Fuentes JC, Pinto Gomes CJ, Cano E. Distribution patterns of endemic flora to define hotspots on Hispaniola. Systematics and Biodiversity. 2016;**14**(3):261-271. DOI: 10.1080/14772000.2015.1135195

[11] Liogier AH. *La Flora de la Española*. Vol. I–IX. Jardín Botánico Nacional Dr. Rafael Ma, Moscoso, Santo Domingo. Dominican Republic: Eastern Central University; 1996-2000. ISBN: 84-8400-217-9

[12] Cano E, Cano OA. Establishment of biogeographic areas by distributing endemic flora and habitats (Dominican Republic, Haiti, R.). In: Biogeography. University Campus, Croatia: Intech Open Access Publisher; 2012. pp. 99-118. ISBN: 978-953-51-0454-4. Available from: https://www.intechopen.com/

[13] Rivas Martínez S, Sánchez Mata D, Costa M. North American Boreal and Western Temperate Forest Vegetation. Itinera Geobotanica. 1999;**12**:5-316

[14] Cano Ortiz A, Musarella CM, Piñar Fuentes JC, Pinto Gomes CJ, del Río S, Cano E. Diversity and conservation status of mangrove communities in two large areas in Central America. Current Science. 2018;**115**(3):534-540

[15] Borhidi A. Phytogeography and Vegetation Ecology of Cuba. Budapest: Akadémiai Kiadó; 1991. 858 pp

[16] Borhidi A. Phytogeography and Vegetation Ecology of Cuba. 2nd ed. Budapest: Akadémiai Kiadó; 1996. 926 pp

[17] Reyes OJ, Acosta CF. Sintáxones de los pinares de *Pinus cubensis* de la zona nororiental de Cuba. Lazaroa. 2012, 2012;**33**:111-169

[18] López Almirall A. Contribución al catálogo de flora cubana: Eendemismos de suelos derivados de ofiolitas. Botánica Complutensis. 2013;**37**:135-152. DOI: 10.5209/rev_BOCM.2013.v37.42276

[19] Cano E, Veloz A, Cano Ortiz A. Phytosociological study of the *Pinus occidentalis* woods in the Dominican Republic. Plant Biosystems–An International Journal Dealing with all Aspects of Plant Biology. 2011;**145**(2):286-297. DOI: 10.1080/11263504.2010.547685

[20] Cano Ortiz A, Musarella CM, Piñar Fuentes JC, Veloz A, Cano E. The dry forest in the Dominican Republic. Plant Biosystems. 2015;**149**(3):451-472. DOI: 10.1080/11263504.2015.1040484

[21] Noguera-Urbano EA. Endemism: differentiation of the concept, methods and applications. Acta Zoológica Mexicana (N S). 2017;**33**(1):89-107

Chapter 2

Salvia ceratophylloides Ard. (Lamiaceae): A Rare Endemic Species of Calabria (Southern Italy)

Giovanni Spampinato, Valentina Lucia AstridLaface, Ana Cano Ortiz, Ricardo Quinto Canas and Carmelo Maria Musarella

Abstract

Salvia ceratophylloides Ard. is a very precious narrow endemism of Southern Italy. It grows in the suburban surroundings of Reggio Calabria, on coastal strip hilly ridges between 250 and 450 m a.s.l. At the beginning of 1900, it was present in several localities, as evidenced by literature, where it was already very rare. Afterward, despite the research carried out by various botanists, the species was no longer found, due to its disappearing in the places mentioned in literature resulting from the intense environmental transformations suffered by the territory. Therefore, the species since 1997 was included in the "Red Book of the flora of Italy" among the extinct species. The successive research carried out in 2008 made it possible to ascertain new localities at about 10 km of distance from those reported in the literature. The actual population consists of about 1000 individuals, and according to IUNC criteria, the conservation status is critically endangered (CR). The threats to survival and spread of the species are different, but above all, it is the habitat destruction due to urbanization to threaten this species.

Keywords: endemism, reggio calabria, critically endangered species, Italy

1. Introduction

The genus *Salvia* has a cosmopolitan distribution and is probably the richest in species of the *Lamiaceae* family. Many species have considerable economic importance for medicinal, food, or ornamental uses. The authors that have dealt with this genus estimate that it contains between 900 and 1000 species [1, 2]. The Mediterranean and Western Asia, with about 240 species, are one of the diversity centers of the genus *Salvia* [2]. The Italian flora includes 20 species of the genus *Salvia* [3] among which *Salvia ceratophylloides* Ard., is a rare endemism of a small hilly area close to the city of Reggio Calabria in Southern Italy (**Figure 1**).

Until 2008, *Salvia ceratophylloides* was considered "extinct in the wild" (EW) by the "Red regional lists of Italian plants" [4], by the "Atlas of endangered species" of Italian flora [5], and by authors who have studied the genus *Salvia* in Italy [6]. The intense urbanization of the areas surrounding the city of Reggio Calabria has led to profound changes in the landscape with the reduction, degradation, and disappearance of natural habitats. All this caused the extinction of the species from the places where it was collected until the beginning of 1900 by various authors [7].

Field surveys carried out in the last decade have, however, allowed us to verify the presence of the species in other localities and to define its taxonomic and ecological characteristics as well as the conservation status [8].

Salvia ceratophylloides currently occurs in small and isolated populations, which are expected to face extinction mainly due to the habitat destruction.

We present a study on the distribution, taxonomy, and conservation status of *S. ceratophylloides*, aimed at identifying the criteria for the management and conservation of this rare species of Italian flora.

Figure 1.
Salvia ceratophylloides *Ard. (A) habit, (B) inflorescence, (C) leaves, and (D) flower.*

2. Materials and methods

We define the distribution of *Salvia ceratophylloides* by several ways such as analysis of herbarium collections, literature data, and through a deep field investigation carried out from 2010 until 2018. To map the distribution, all sites of species were geo-referenced in the field and the coordinates archived in a geographical information system (GIS).

The morphological traits were measured on 10 herbarium specimens with a stereomicroscope. The specimens are kept at the "Mediterranea" University of Reggio Calabria Herbarium (REGGIO). To assess population size and the habitat, several surveys were carried out on the *S. ceratophylloides* biotope. Conservation status was assessed according to the IUCN Red List criteria [9], which is actually the more widely used in the world to assess extinction risk. The nomenclature of the species is according to [10].

3. Historical vicissitudes

The first news on this species dates back to the pre-Linnaean period and is due to Cupani [11], a Franciscan monk, an expert on Sicilian flora, who in his monumental work "*Panphyton siculum*" published the iconography and description of many species of Sicilian flora. The work published posthumously by the publisher

Figure 2.
Iconography from the "Panthyton Siculum" by Cupani of 1713 related to Salvia ceratophylloides.

in 1713 [12] contains an iconography (t 186) clearly referable to *Salvia ceratophylloides* (**Figure 2**) that Cupani describes with the phrase "*Sclarea verbenae petalis purpurates florae.*" Cupani described many species in his book that he collected spontaneously in Sicily, but also non-native, grown at the botanical garden of Misilmeri in Sicily, of which he was the curator [13]. This has created confusion in Sicilian botanists who for a long time have attributed this species also to the Sicilian flora [14].

The species was described with the binomial system by Pietro Arduino (1728–1805) [15] as *Salvia ceratophylloides*, in "*Animadversionum botanicarum specimen alterum*" (**Figure 3**).

Arduino for the description used plants grown at the Botanical Garden of Padua, obtained from seeds that received from a friend from Sicily as he says "*Ex seminibus e Sicilia ab amico delatis, et in horto hoc nostro satis quinque anni sab hinc nata est.*" Also for this reason, Sicily was considered for a long time the place of origin of this species, although no one has ever collected samples in the wild. Arduino attributed to this sage the epithet "*ceratophylloides*" to highlight the similarity that

Figure 3.
Iconography of Salvia ceratophyllales *from Arduino [15].*

in his opinion, this species has with *Salvia ceratophylla* L., a species distributed in the Irano-Turanian region absent in Italy (from the Greek εἶδος eidos appearance, resemblance). Arduino describes the species as: "*Salvia ceratophylla sed luxuriantis similis; forte diversa ac distincta*").

Arduino deposited the type specimens used for the species description, in its herbarium of Vicenza, which unfortunately was destroyed during the Second World War. He sent some herbarium samples to Linnaeus in 1763, still preserved at the Linnaean herbarium at London. The sample sent and inserted by Linnaeus in his herbarium No. 42.55 (**Figure 4**) shows the name of the species and connected to a list attached to some *exsiccata* sent by Arduino to Linnaeus.

Lacaita [7] organized the taxonomic framework of *Salvia ceratophylloides*, clarifying the real place of origin of the species. He pointed out that the species has never been found spontaneously in Sicily and that the few specimens present in various herbaria are all coming from southern Calabria or from crops obtained by seeds always coming from these localities.

Lacaita [16] found that the samples named by Linnaeus as *Salvia ceratophylla* L. and by subsequent authors as *S. ceratophylloides* L. are, in fact, to be attributed to *Salvia ceratophylloides* Ard. Therefore, *S. ceratophylloides* L. is a synonym of *S. ceratophylloides* Ard., which has a nomenclatural priority.

Figure 4.
The herbarium specimens of Salvia ceratophylloides *sent by Arduino to Linnaeus in 1763 and stored in the Linnaean herbarium at London (LINN 42.55).*

Until the beginning of 1900, *Salvia ceratophylloides* was collected by various botanists [7, 17, 18] from some locations around Reggio Calabria: Gallico Superiore, Terreti, Straorino, Ortì, Vito Superiore, Pietrastorta, even then the species was very rare [7]. In the following decades, the species was not found again despite the research carried out by various botanists, as the profound environmental transformations due to the urban expansion of the city of Reggio Calabria determined the destruction of the habitat and disappearance of the species. In 1997, the species was considered extinct and included in the "Red Book of the flora of Italy" as extinct species (EX) [4–6, 19]. Field research started in 2008, however, allowed to discover some small new populations of *S. ceratophylloides* around Reggio Calabria (Puzzi, Cataforio) [20]. Later were found other small populations in places hardly accessible, always in the surroundings of Reggio Calabria [21].

4. Morphological trait

Salvia ceratophylloides (**Figures 1**, **3** and **4**) is a perennial herbaceous plant (scapose hemicryptophyte), densely pubescent for both glandular and simple patent hairs. The plant is 30–90 cm tall, has upright or ascending stems, normally lignified and much ramified at the base. The leaves are pinnate-partite, with toothed lobes, the basal of 10–12 × 4–6 cm, and the cauline of 3–4 × 1–2 cm. The leaf blade, wrinkled on both sides, is glandular and pubescent, more evident on the lower one. The inflorescences are very showy, 20–30 cm long, normally ramified at the base, and are made up of 5–6 verticillaster each with 4–6 flowers. The calyx, 8–10 × 3–5 mm, is zygomorphic, ferruginous, and covered with numerous sessile or pedicel glandular hairs. The bilabiate corolla, violet, 15–25 mm long (three times the calyx) has an upper lip strongly folded to the cap on the stamens pubescent-glandular on the outside. The two stamens are 10 mm long and the style 20–30 mm long, protrudes over 10 mm from the upper lip and has a deeply bifurcated purple-colored stigma at the extremity. The fruit is a peculiar schizocarp: a microbasarium made up of four dark brown, spherical to ovoidal mericarps; each mericarp, 2–3 mm long, has a thickened margin.

Salvia ceratophylloides has a main flowering period in spring from April to June and a second flowering period in autumn from October to November. The fruiting occurs after some flowering weeks.

It is a strongly aromatic plant, rich in essential oils [22], and of future potential economic interest. The composition of volatile substances produced by glandular hairs suggests that these substances are involved in chemical defense against insects.

5. Taxonomic remark

The taxonomic rank of *S. ceratophylloides* is controversial. Hedge [23] considers it a synonym of *Salvia pratensis* subsp. *pratensis*, instead the greater number of the authors indicate it as a good species [19, 24]. The species is reported with different taxonomic classification in the Italian floras: Fiori & Paoletti [25] reported *S. ceratophylloides* as a variety of *S. pratensis*, Tenore [17] and Pignatti [3], considering it as an autonomous species. The most recent checklists of the Italian flora [10] and the authors who dealt with the taxonomy of the *Salvia* genus in Italy [6] also consider *S. ceratophylloides* an autonomous endemic species of Southern Calabria.

Salvia ceratophylloides Arduino Animad. Bot. Spec. Alt. 12, tab 2.1764.

Synonyms: *Salvia ceratophylloides* L., Mantissa Pl., 26, 1767; *Salvia pratensis* L. var. *ceratophylloides* (Ard.) Fiori e Paoletti, Flora Analitica d'Italia, 5: 255, 1909; and *Salvia ceratophylloides* Ten. var. *A* Ten., Flora Napolitana 3: 25, 1829.

	Plant	Lamina of the cauline leaves	Inflorescence	Corolla
S. ceratophylloides	30–90 cm high, pubescent and very glandular	Pinnate-partite with serrated lobes	Wide, very branched	15–25 mm long with numerous glandular hairs
S. pratensis subsp. *pratensis*	30–60 cm high pubescent and little or not glandular	Ovate-oblong with slightly crenate or incised lobed margin	Cylindrical, little, or not branched	13–20 mm long without or with few glandular hairs
S pratensis subsp. *saccardiana*	50–110 cm high, pubescent, and glandular	Cordate-ovate, with lobed-crenate margin	Cylindrical, little, or not branched	28–25 mm long, with numerous glandular hairs
S. haematodes	50–120 cm high, pubescent and very glandular	Triangular, with crenate margin	Wide, very branched	28–35 mm with numerous glandular hairs
S. virgata	30–100 cm high, pubescent and glandular especially in the upper part	Ovate-oblong with slightly crenate or incised lobed margin	Wide, very branched with arcuate-erect branches	13–16 mm glandular only at the margin

Table 1.
Main morphological differences between species of Salvia pratensis *group in Italy.*

Salvia ceratophylloides belongs to Subgenus *Sclarea* (Moench) Benth., Section *Plethiosphace* Benth., characterized by a campanulate calyx, above two furrowed, concave, with three conniving teeth and corolla with a crescent-shaped upper lip and compressed. It is a part to the group of *S. pratensis* present in Italy with: *S. pratensis* L. subsp. *pratensis*, *S. pratensis* L. subsp. *saccardina* (Pamp.) Poldini, *S. haematodes* L., *S. virgata* Jacq., and *S. ceratophylloides* Ard. [3]. *S. ceratophylloides* is well distinguished from the other sages species of the *Salvia pratensis* group, above all for the pinnate-partite leaf with toothed lobes. **Table 1** shows the main morphological differences between the species of *Salvia pratensis* group in Italy.

The type of *Salvia ceratophylloides* was lost because the Arduino herbarium was destroyed during the Second World War. Del Carratore et al. [26] carried out the lectotyping of *S. ceratophylloides* Ard. Among all the samples, present in the various legally recognized herbarium, they are indicated as the lectotype of *Salvia ceratophylloides*, the specimen that Arduino sent to Linnaeus stored in the Linnaean Herbarium of London at N° 42.55 (**Figure 4**).

6. Habitat

Salvia ceratophylloides grows near Reggio Calabria, on the hills of the western side of the Aspromonte facing the Strait of Messina (Southern Italy), at altitudes between 250 and 400 m a.s.l. especially on the cooler slopes with the north or northwest exposition. The geology of the area is characterized by layers of loose sand alternating with benches of soft calcarenites of Pliocene origin. The soils have a sandy texture with a basic pH and fall into the group Calcaric Cambisols [27].

The area has a typically Mediterranean climate with average annual temperatures of 18°C and average annual rainfall of 600 mm, concentrated in the autumn

Figure 5.
Habitat of Salvia ceratophylloides.

months of November and December, and a summer dry period of about 5 months. According to Rivas Martinez et al. [28], the macro-bioclimate is "Mediterranean pluviseasonal oceanic, with thermo Mediterranean bioclimate" (upper thermo Mediterranean thermotype and a lower subhumid ombrotype).

Salvia ceratophylloides grows spontaneously in the habitat of the EEC directive 43/93:"5330 thermo Mediterranean and predesert scrub" subtype "32.23 Diss-dominated garrigues." This habitat includes Mediterranean steppe and garrigue dominated by *Ampelodesmos mauritanicus* (Poir.) Dur. & Schinz. or *Artemisia campestris* L. subsp. *variabilis* (Ten.) Greuter (**Figure 5**) [29, 30]. The most frequently growing species with *S. ceratophylloides*, in addition to the aforementioned species, are some grasses (*Lagurus ovatus* L., *Avena barbata* Link, *Briza maxima* L., *Hyparrhenia hirta* (L.) Stapf., *Dasypyrum villosum* (L.) P. Candargy), several dwarf shrubs (*Cistus creticus* L., *Cistus salviifolius* L., *Micromeria graeca* (L.) Benth. ex Rchb., *Thymbra capitata* (L.) Cav., *Phlomis fruticosa* L.), and some shrubs (*Cytisus infestus* (C.Presl) Guss. subsp. *infestus*, *Spartium junceum* L.). Mostly, they are widespread species in the steppic grassland and in the Mediterranean garrigues.

7. Distribution and abundance of the population

The field researches conducted in 2008–2010 allowed discovering four small new populations of *S. ceratophylloides* around Reggio Calabria (Puzzi, Cataforio). Each population consisted of a few dozen individuals, for a total census of 100 mature individuals [20, 31]. After looking everywhere, in recent years, other eight small populations of *S. ceratophylloides* in places hardly accessible were found, always in the surroundings of Reggio Calabria (**Figure 6**). The total area occupied by the 12 sites is about 4200 m^2. In these sites, the plants are distributed as follows: 601 breeding without flowers or fruits, 22 flowering individuals, 3 fruiting individuals, 216 seedlings, 259 juvenile individuals nonbreeding, and 17 dead individuals. For a total of about 1100 individuals [21]. The distribution of plants in the fragmented subpopulations is very irregular in relation to the quality of the habitat. In some sites, the number of plants is more abundant, with about 250 individuals; in other sites, the individuals do not exceed 3 units.

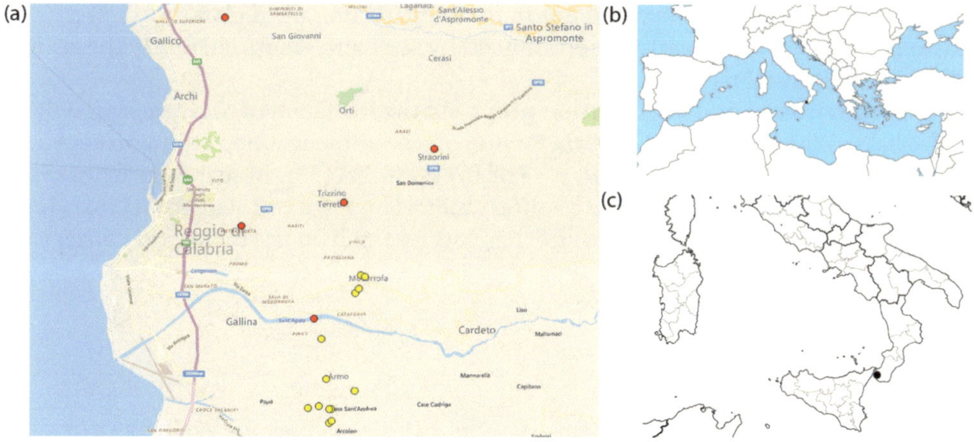

Figure 6.
(a) Distribution map of Salvia ceratophylloides: *red dot—populations cited in the literature and extinct; yellow dot-population currently existing found after 2008. (b) Mediterranean Basin: black dot-distribution area and (c) Southern Italy.*

8. Conservation status

As a consequence of population disappearance from the historical places where *Salvia ceratophylloides* was collected at the beginning of the 1900s, it was considered extinct globally (EX) [4, 5]. In 2010, after the discovery of some sites, an assessment was made from Crisafulli et al. [20], and according to the IUCN criteria, the species was ascribed to the "critically endangered" risk categories (CR). Research carried out more recently by Laface et al. [21] ascertained an increasing population size in number of breeding adults and in the area of occupied populations. Nevertheless, the new assessment, using the IUCN criteria B, C, and D, confirmed the status of "critically endangered" species. So, *S. ceratophylloides* is seriously threatened, facing an immediate high risk of extinction.

S. ceratophylloides has a naturally small population and therefore susceptible to damaging influences. The remarkable fragmentation of the habitat, also due to the environmental transformations caused mainly by various anthropic activities, is the major conservation problem.

According to EIONET classification [32], the threats that influence the survival of this sage can be codified as follows: F01: conversion from other land uses to housing, settlement or recreational areas (excluding drainage and modification of coastline, estuary, and coastal conditions); A02: conversion from one type of agricultural land use to another (excluding drainage and burning); A03: conversion from mixed farming and agroforestry systems to specialized (e.g., single crop) production; A15: tillage practices (e.g., plowing) in agriculture; A16: other soil management practices in agriculture; A17: harvesting of crops and cutting of croplands; B01: conversion to forest from other land uses, or afforestation (excluding drainage); C01: extraction of minerals (e.g., rock, metal ores, gravel, sand, shell); A11: burning for agriculture; A13: burning for forestry; I02: other invasive alien species (other than species of Union concern); and L06: interspecific faunal and floral relations (competition, predation, parasitism, pathogens). The latter threat, due to insects that feed on fruits and flowering stems, is particularly serious and strongly compromises the reproductive capacity of the species.

Salvia ceratophylloides currently is not safeguarded *in situ* from protected areas or from Sites of Community Importance (SCI) according to Directive CEE 43/92.

The *ex situ* conservation is carried out in the Messina Botanical Garden "Pietro Castelli," where it is successfully reproduced by seed and where currently there are about thirty individuals.

Some environmental organization proposed the inclusion of *S. ceratophylloides* in the lists to the laws of the Calabria Region on the autochthonous flora protection (L.R. n.30 of 26/11/2001; L.R. n.47 of 07/12/2009) [33]. Researchers from the Department of Agriculture of the Mediterranean University of Reggio Calabria have initiated, on its own initiative, monitoring and research on the biological conservation of this rare sage [21].

9. Conclusion

Endemic species have a high conservation priority, as they are exclusive to a geographically restricted and limited territory. Being often rare and ecologically infrequent, any unfavorable change can cause their rapid extinction [34]. The exclusive endemisms deserve the conservation priority since the disappearance of the local populations caused the complete extinction of these species. Considering the regional responsibility for the conservation of species [35] and that *S. ceratophylloides* is one of the most endangered rare endemic plants, this species should be the subject of specific *in situ* and *ex situ* conservation actions.

The involvement of local authorities (Calabria region and metropolitan city of Reggio Calabria), responsible for the conservation of natural resources and the environment, is essential to take concrete actions aimed at the conservation of this species.

To protect populations of this species and preserve them over time, the establishment of microreserves, following the example applied in Spain [36], could be a good solution for the *in situ* conservation of this rare sage. These are small protected areas created in order to ensure the future study and monitoring of the endangered endemic flora plants, which can be entrusted to environmentalist associations or to the landowner.

Author details

Giovanni Spampinato[1*], Valentina Lucia Astrid Laface[1], Ana Cano Ortiz[2], Ricardo Quinto Canas[3,4] and Carmelo Maria Musarella[1,2]

1 Department of Agraria, "Mediterranea" University of Reggio Calabria, Reggio Calabria, Italy

2 Department of Animal Biology, Plant Biology and Ecology, Jaen University, Jaén, Spain

3 Faculty of Sciences and Technology, University of Algarve, Faro, Portugal

4 Centre of Marine Sciences (CCMAR), University of Algarve, Faro, Portugal

*Address all correspondence to: gspampinato@unirc.it

References

[1] Kintzios SE. Sage—The Genus *Salvia*. Boca Raton: CRC Press; 2003. 318 p

[2] Walker JB, Sytsma KJ, Treutlein J, Wink M. *Salvia* (Lamiaceae) is not monophyletic: Implications for the systematics, radiation, and ecological specializations of *Salvia* and tribe Mentheae. American Journal of Botany. 2004;**91**:1115-1125. DOI: 10.3732/ajb.91.7.1115

[3] Pignatti S. Flora d'Italia 2 edizione. Bologna: Edagricole; 2018;**3**:301-310

[4] Conti F, Manzi A, Pedrotti F. Liste Rosse Regionali delle Piante d'Italia. Camerino: Assoc. Ital. WWF, Società Botanica Italiana; 1997

[5] Scoppola A, Spampinato G. Atlante delle specie a rischio di estinzione. Versione 1.0. CD-Rom enclosed to the volume. In: Scoppola A, Blasi C, editors. Stato Delle Conoscenze Sulla Flora Vascolare d'Italia. Roma: Palombi Editori; 2005

[6] Del Carratore F, Garbari F. Indagini Biosistematiche Sul Genere *Salvia* L. Sect. *Plethiosphace* Benyham (*Labiatae*). Informatore Botanico Italiano. 1997;**29**:297-299

[7] Lacaita C. Addenda et emendanda ad floram italicam. Bullenttino della Societa Botanica Italiana. 1921;**28**:18-19

[8] Spampinato G, Crisafulli A, Marino A, Signorino G. *Salvia ceratophylloides* Ard. Informatore Botanico Italiano. 2011;**43**(2):381-458

[9] IUCN. The IUCN Red List of Threatened Species. Version 2015-4. 2018. Available from: http://www.iucnredlist.org [Accessed: November 30, 2018]

[10] Bartolucci F, Peruzzi L, Galasso G, Albano A, Alessandrini A, Ardenghi NMG, et al. An updated checklist of the vascular flora native to Italy. Plant Biosystems–An International Journal Dealing with all Aspects of Plant Biology. 2018;**152**(2):179-303

[11] Cupani F. Pamphyton Siculum. Panormi: Epiro; 1713. 1440 pp

[12] Costa RMS, Pavone P, Carbonaro RA, Pulvirenti S. Francesco Cupani's panphyton siculum. The rediscovery of a copy with handwritten corrections by his pupil Antonio Bonanno. Botany Letters. 2016;**163**(2):203-216. DOI: 10.1080/23818107.2016.1166983

[13] Proietto P. Francesco Cupani–Naturalista Contemporaneo. Palermo: Lulu.com; 2013. ISBN-13: 978-1291626650

[14] Gussone G. Floræ Siculæ Prodromus. Vol. I. Napoli; 1827. p. 19

[15] Arduino P. Animadversiorum Botanicorum Specimen Alterum. Ex Typographia Sansoniana: Venetis; 1764

[16] Lacaita C. Piante italiane critiche o rare: 67. *Salvia ceratophylloides* Arduino. Nuovo Giornale Botanico Italiano. 1921;**28**:144-147

[17] Tenore M. Sylloge plantarum vascularium florae neapolitanae hucusque detectarum. In: Flora Napolitana 3. Neapoli: Ex Typographia Fibreni; 1831

[18] Macchiati L. Catalogo delle piante raccolte nei dintorni di reggio calabria dal Settembre 1881 al Febbraio 1883. Nuovo Giornale Botanico Italiano. 1884;**16**:59-100

[19] Del Carratore F, Garbari F. Il gen. *Salvia* sect. *Plethiosphace* (Lamiaceae) in Italia. Archivio Geobotanico. 2001;**7**(1):41-62

[20] Crisafulli A, Cannavò S, Maiorca G, Musarella CM, Signorino G, Spampinato

G. Aggiornamenti floristici per la Calabria. Informatore Botanico Italiano. 2010;**42**(2):431-442

[21] Laface VLA, Musarella CM, Spampinato G. Conservation status of the Aspromontana flora: Monitoring and new stations of *Salvia ceratophylloides* Ard. (*Lamiaceae*) endemic species in Reggio Calabria (Southern Italy). In Book of Abstract–113°Congresso della Società Botanica Italiana. V International Plant Science Conference (IPSC); 12-15 September 2018; Fisciano (SA). p. 96. ISBN: 978-88-85915-22-0

[22] Spampinato G, Araniti F, Vescio R, Musarella CM, Di Iorio A, Abenavoli MR, et al. VOCs composition of *Salvia ceratophylloides* Ard. (*Lamiaceae*), a Calabrian endangered species, through targeted and untargeted metabolomic analyses. In Book of Abstract–113° Congresso della Società Botanica Italiana. V International Plant Science Conference (IPSC);12-15 September 2018; Fisciano (SA). p. 33. ISBN: 978-88-85915-22-0

[23] Hedge IC, Salvia L. In: Tutin TG, Heywood VH, Burges NA, Moore DM, Valentine DH, Walters SM, Webb DA, editors. Flora Europaea. Vol. 3. Cambridge: Cambridge University press; 1972. pp. 188-192

[24] Greuter W, Burdet H, Long G. Med-Checklist. A Critical Inventory of Vascular Plants of the Circum-Mediterranean Countries. Vol. 3. Conservatoire et Jardin Botaniques de la Ville de Genève: OPTIMA; 1986

[25] Fiori A, Paoletti G, editors. Flora analitica d'Italia, Vol. III. Labiate (CXIV), 1. Padova: Tipografia del seminario; 1903-1904

[26] Del Carratore F, Garbari F, Jarvis C. Typification of the names of three mediterranean sage species: Salvia virgata Jacq., *S. Ceratophylloides* Ard. and S. Bertolonii Vis. (*Lamiaceae*). Allionia. 1999;**36**:41-46

[27] FAO-Unesco. Soil Map of the World, Revised legend. Rome: FAO; 1990

[28] Rivas-Martinez S. Global Bioclimatics (Clasificacion Bioclimatica de la Tierra). Phytosociological Research Center. Available from http://www.globalbioclimatics.org/book/bioc/global_bioclimatics-2008_00.htm [Retrieved: December 8, 2018]

[29] Brullo S, Scelsi F, Spampinato G. La Vegetazione dell'Aspromonte. Studio Fitosociologico. Reggio Calabria: Laruffa Editore; 2001. pp. 1-370. ISBN 88-7221-160-3

[30] Spampinato G, Cameriere P, Caridi D, Crisafulli A. Carta della biodiversità vegetale del Parco Nazionale dell'Aspromonte. (Italia Meridionale). Quaderni di Botanica Ambientale e Applicata. 2008;**19**(3-36):8-12

[31] Spampinato G, Crisafulli A. Struttura delle popolazioni e sinecologia di *Salvia ceratophylloides* (Lamiaceae) specie endemica minacciata di estinzione. Atti 103° congr. S.B.I.; 17-19 sett. 2008; Reggio Calabria. p. 56

[32] EIONET. 2018. Available from: http://biodiversity.eionet.europa.eu/activities/Reporting/Article_17/Reports_2019/Files_2019/Pressures_Threats_Final_20180507.xls [Accessed: December 12, 2018]

[33] Spampinato G. Guida Alla Flora dell'Aspromonte. Reggio Calabria: Laruffa Editore; 2014. ISBN 978-88-7221-707-8

[34] Callmander MW, Schatz GE, Lowry PP II. IUCN red list assessment and the global strategy for plant conservation: Taxonomists must act now. Taxon. 2008;**54**(4):1047-1050

[35] Gauthier P, Debussche M, Thompson JD. Regional priority setting

for rare species based on a method combining three criteria. Biological Conservation. 2010;**143**:1501-1509

[36] Laguna E. The Micro-reserves as a Tool for Conservation of Threatened Plants in Europe. Strasbourg: Nature and Environment series no 121. Council of Europe; 2001. 119 pp

Chapter 3

Endemic Plant Species of Bolivia and Their Relationships with Vegetation

Mónica Moraes R., Carla Maldonado and Freddy S. Zenteno-Ruiz

Abstract

The inventory of Bolivia's vascular plants lists 2402 endemic species (ca. 20% of 12,339 of native flora). Among angiosperms, there are 2263 species from 124 families and 641 genera, whereas among pteridophytes, there are 139 species from 16 families and 29 genera. Seven families with the greatest number of endemic species are Orchidaceae (418), Asteraceae (246), Bromeliaceae (147), Cactaceae (127), Poaceae (92), and Piperaceae (81). *Cleistocactus* and *Puya* have 14 and 55 endemic species, respectively, so representing 82.3 and 84.6% of the species in these genera. Bolivia's endemic species show distribution patterns associated with past geological events, orographic dynamics (of the Andes and in the Cerrado), as well as areas of diversification. Dry xeric and humid regions host local and regional endemics in specific families and biogeographic regions of high conservation importance. Humid montane forests in the Yungas and dry inter-Andean valleys are rich in endemic species with 51 and 22% of the total recorded in the respective regions. Nevertheless, there are still many lesser known geographical areas that may generate new information in the short and medium term. Only 165 endemic species (6.9%) have been evaluated for their conservation status following IUCN categories with 49% assessed as endangered (EN).

Keywords: angiosperms, humid montane forests, ferns, dry valleys, Cerrado

1. Introduction

The Bolivian biota and its endemicity are derived from the influence of four biogeographic provinces, the Amazonian, the Andes, the Gran Chaco, and the Cerrado, generating several encounters of mixed elements [1]. For example, in the mountains of the Eastern Cordillera with the mixture of Andean and Amazonian flora, while in the Pantanal area (SE Bolivia) where there are Amazonian, Chaco and Cerrado elements took place. Amazonia is found in the alluvial plain from center to the north of the country, the Andes in mountain ranges in the west side, Cerrado on Precambrian shield in the east (in which it is also circumscribed to the Chiquitanía), and the Gran Chaco on plains and Andean foothills in the south.

IntechOpen

These four biogeographical provinces together with the physiography of Bolivia combine in general opportunities for isolation, speciation, and restricted distribution, especially in geological periods, such as the uplift of the Andes and the conformation of valleys [1–3]. The floristic elements of each biogeographic province derive from radiation and dispersion processes during geological ages, whose adaptation has been consolidated in current landscapes.

Endemic species in Bolivia are associated with a wide range of diverse habitats and originated from different processes that have modeled the natural landscape in the past, particularly in the montane formations lying on the Andean and Brazilian lithospheres, specifically the uplift of the central Andes [4] 18 million years ago in the west and the slow weathering of the Precambrian Brazilian shield in the east, for example, through three biogeographical models of speciation in central Andes: vicariance of Andean uplifts, dispersal during the Pleistocene, and vicariance in both eastern and western slopes during glacial periods [5]. The rocky outcrops exposed in the Cerrado (or "brasileño-paranaense") of Bolivia have been identified as centers of plant endemism, where microhabitats are formed with specific microclimatic conditions [6, 7].

In the present contribution, we update the record of endemic vascular plants of Bolivia and evaluate general striking features and patterns of endemic plants and their relationships with vegetation types of the country, as a baseline for further studies.

2. Methodology

A large part of the information collected in the country is concentrated in databases and botanical collections of the National Herbarium of Bolivia (acronym, LPB in the Index Herbariorum). Due to a constant communication with experts and specialists from different groups of plants, publications and other valid and current scientific denominations about recent taxonomic works are available. This support constitutes our main source of references. Recently, an intense data compilation process has resulted from the information gathered for the "Catálogo de las Plantas Vasculares de Bolivia," published in 2014 [8] which mainly continues to be updated in Tropicos (www.tropicos.org, revised until November 7, 2018), as well as in other recent publications. The botanical inventory of Bolivia has been documented based on 650,000 herbarium specimens. The distribution of the collections covers a higher density in regions of the eastern slopes of the Andes, which includes the humid forests and the fragmented forests of the dry inter-Andean valleys (**Figure 1**). Recent publications and taxonomic updates to the Bolivian flora of vascular plants were the basis for this work [9–17].

In the highlands of the Andes toward the southwest and also in ravines and hills of the eastern Andean mountains, *Polylepis* (Rosaceae) forests (**Figure 2a**) are found, which are considered to be the highest in the world. However, they have been excluded here as a major vegetation formation because they are distributed in both a fragmented and azonal pattern in different vegetation formations, such as montane forests, dry puna, and humid puna. Therefore, a slightly modified map according to two vegetational classifications [8, 18] was elaborated (**Figure 3**).

The consideration of the endemic species follows in relation to the vegetation formations in which they are represented, which allows an evaluation on the distribution and striking characteristics of the families represented. This information is

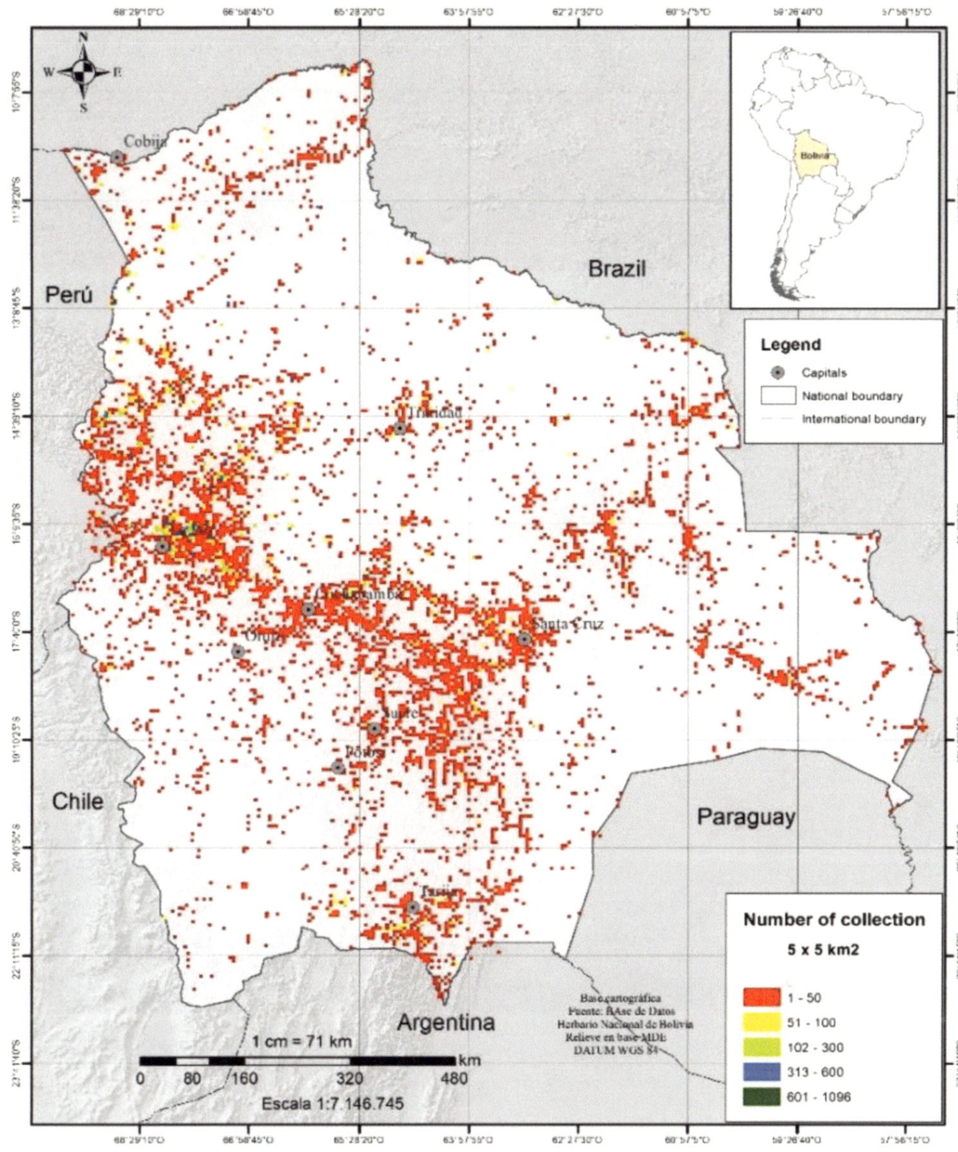

Figure 1.
Density cover of botanical collections in Bolivia at LPB (grid 5 × 5 km).

subsequently taken care of according to the biogeographical implications of the Bolivian flora. Finally, a number not exceeding to 200 endemic species occur in uncertain vegetation formations.

According to the Bolivian plant checklist [8], there are four major geographical regions in Bolivia: the Andes, Altiplano, lowlands, and uncertain. However this is an arbitrary delimitation that does not recognize local differences. The Andes, for example, has an altitude range of 500–3500 m and includes the Yungueñan "páramo" and both humid and dry forests in the Andean foothills. The Altiplano lies above 3500 m, whereas the lowlands include Amazonian rainforests, savannas, the Pantanal, and the Chaco xeric forest from 150 to 400 m, as well as the Precambrian mountain ranges of 600–1200 m altitude.

Figure 2.
Vegetation formations of Bolivia. (a) Polylepis *forest, (b) humid forest, (c) the Yungas, (d) Tucumanian-Bolivian forests, (e) "Campos cerrados," and (f) Chaqueñan foothills forests.*

The conservation status of endemic Bolivian plants has been evaluated in the four published red books [19–22] focused on plants of the Cerrado, wild relatives of cultivated plants, threatened plants of the Andean zone, and threatened plants of the lowlands.

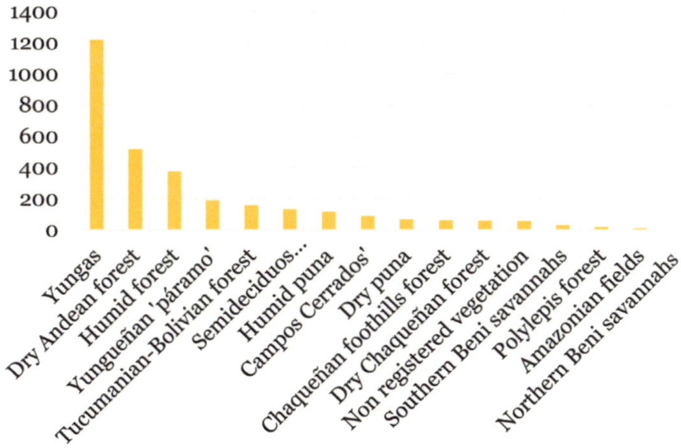

Figure 3.
Vegetation formations of Bolivia (modified from [18, 8]).

3. Results: what do we know to present?

3.1 Diversity and distribution

The endemic floristic elements represent biogeographic provinces that are found in Bolivia (**Table 1**). The highest representativeness comes from the Andes with 81% for the endemic plants of Bolivia, followed by the Amazon (16%), the Cerrado (9%), and the Gran Chaco (3%), whereas in four reported natural regions registered for each specimen in www.tropicos.com, trends in endemism are shown: the Andes (69.4%), lowlands (22.3%), Altiplano (6.6%), and uncertain (1.65%).

Although there are many mixtures of floristic elements in the country, it is generally recognized that certain vegetation formations are related to biogeographical affinities. The Amazonian province includes in general four vegetation types: northern and southern Beni savannahs, humid forest, and Amazonian camps, ranging from 100 to 400 m elevation. Both dry and humid "punas," *Polylepis* forests, dry inter-Andean valleys, Tucumanian-Bolivian forest, the Yungas, and the Yungueñan "páramo" make up the Andean province with 1500–5500 m elevation. For the Cerrado, the vegetation formations of the "campos cerrados" and the semideciduous Chiquitanian forest are included, between 400 and 1100 m elevation. And finally, both dry Chaqueñan forest and Chaqueñan foothills forest belong

	Floristic elements
Andean province	*Aa* (5), *Adesmia* (1), *Aspidosperma* (1), *Baccharis* (13), *Capsicum* (3), *Catasetum* (8), *Croton* (4), *Festuca* (17), *Nototriche* (8), *Passiflora* (22), *Protium* (1), *Solanum* (20), *Stevia* (25)
Amazonian province	*Annona* (1), *Andropogon* (3), *Attalea* (1), *Axonopus* (2), *Diospyros* (2), *Ficus* (1), *Heliconia* (1), *Machaerium* (3), *Nectandra* (2), *Neea* (5), *Paspalum* (4), *Sloanea* (1)
Cerrado province	*Acosium* (1), *Arachis* (7), *Borreria* (4), *Bromelia* (1), *Calea* (3), *Cordia* (1), *Discocactus* (2), *Frailea* (2), *Oxypetalum* (4), *Syagrus* (1), *Vellozia* (1)
Gran Chaco province	*Bulnesia* (1), *Cereus* (3), *Cnidoscolus* (2), *Gaya* (3), *Izozogia* (1), *Pereskia* (3), *Portulaca* (2)

Table 1.
Some characteristic genera of the four main biogeographic provinces in Bolivia. The number of endemic species is indicated in brackets.

to the Gran Chaco at 400–700 m. The meeting point of biogeographic Amazonian and Andean elements is mostly represented in the Yungas of the eastern slopes of the Andes, between 1500 and 3000 m elevation.

The endemic species of Bolivia make up a diverse group of life forms and taxonomic groups that contribute to the diversity of natural landscapes. Among these are trees (Annonaceae, Arecaceae, Erythroxylaceae, Fabaceae, and Lauraceae), tree ferns (Cyatheaceae), shrubs (Melastomataceae, Piperaceae), and subshrubs (Asteraceae and Ericaceae), as well as vines (Passifloraceae) and succulents (Cactaceae). There are also numerous herbaceous plants (Cyperaceae, Gesneriaceae, Iridaceae, and Poaceae), forbs (Heliconiaceae and Marantaceae), prostrate and scandent herbs (Aristolochiaceae, Convolvulaceae, and Cucurbitaceae), ferns (Dryopteridaceae and Polypodiaceae), aquatics (Eriocaulaceae and Isoëtaceae), and epiphytes (Loranthaceae and Orchidaceae).

Endemic plants were recorded from all 14 vegetation types known from Bolivia (see some examples in **Figure 2**). In terms of the richness of endemic species found in different vegetation formations, the trend is similar to that of plant families. The vegetation formation of the Yungas (humid mountains of the eastern Andes) hosts the highest concentration of endemic plant with 1218 (51% of the total number of endemic plants, **Figure 4**); it is followed by the dry inter-Andean valleys with 518 species and humid forests with 375. There are less than 50 endemic species found in vegetation formations in the alluvial plains of the lowlands (northern and southern Beni savannahs and "campos amazónicos"). Below is a synopsis of a sample on six vegetation formations represented in Bolivia with the largest number of endemic species recorded in relation to the ten families with highest level of endemism listed in in **Figure 5**. Most endemic species are restricted 1(−2) to a specific vegetation formation, but 1.26% are distributed in four or more contiguous formations (**Table 2**).

The Yungas: It is located along the eastern slopes of the Andean mountain ranging between 1000 m and 3500 m. It includes the timberline at higher elevations where it forms the Yungueñan "páramo" or cloud and humid montane forests at middle altitudes and sub-Andean and lowland forests at lower altitudes. The height of the forests decreases with increasing elevation, and, especially in cloud forests, the trees are covered with mosses and other epiphytic plants (**Figure 2c**). The diversity of tree species is higher at low elevations and decreases as altitude increases. Bamboos (*Chusquea* spp.) and tree ferns (*Cyathea* spp.) are also frequently found here.

A total of 139 families are found in the Yungas (**Figure 5a**). Orchidaceae is the most abundant in species numbers (with 294), followed by Asteraceae (130) and Piperaceae (51). Among the more important genera are *Lepanthes* (Orchidaceae) with 50 species, *Peperomia* (Piperaceae) with 35, and *Elaphoglossum* (Dryopteridaceae) with 29.

Dry Andean forest: The vegetation is often deciduous as a result of the long dry season. Most of the native vegetation occurs in isolated populations due to an intensive agricultural expansion or the widespread plantation of exotic species, such as *Eucalyptus* spp., *Phragmites* spp., and *Pinus radiata* D. Don.

Of more than 3000 species recorded for this region, almost 16% are endemic. A total of 139 families are represented with Asteraceae having the highest number of species (97) followed by Cactaceae (83) and Bromeliaceae (52) (**Figure 5b**). Among the most important genera are *Stevia* (Asteraceae), *Tillandsia* (Bromeliaceae) with 24 species, and *Puya* (Bromeliaceae) with 20.

Humid forest: It mostly covers the alluvial plains in the lowlands of Bolivia. The topography is relatively flat from about 500 m at the lower limits of the Yungas forests down to 100 m elevation in the north, mostly evergreen with continuous

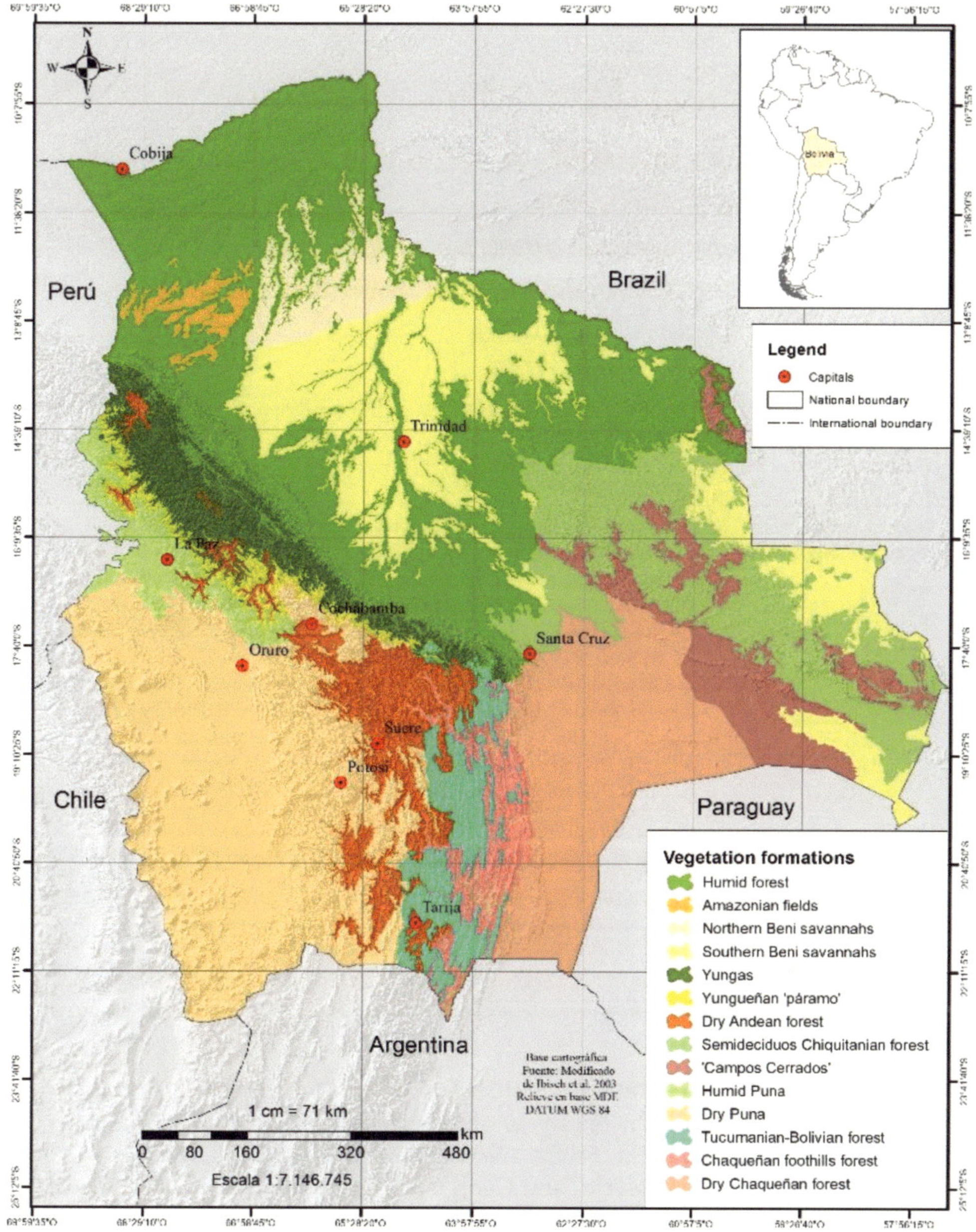

Figure 4.
Distribution of a number of endemic plants of Bolivia in major vegetation types, including highlands and lowlands.

tree canopy and characterized by their high diversity (**Figure 2b**). Associations of monotypic or impoverished forests dominated by large bamboos (*Guadua* spp.) or palm trees (*Mauritia flexuosa* L.f., *Oenocarpus bataua* Mart.) that successfully suppress the growth of other species are also found here.

Of the 5663 species recorded, almost 6% are endemic coming from 138 families. Orchidaceae is the most important family (with 57 species) followed by Piperaceae (35) and Bromeliaceae (24) (**Figure 5c**). *Piper* and *Peperomia* (Piperaceae) are the genera with more endemic species, 18 and 17 species, respectively. *Fosterella* (Bromeliaceae) is another genus with several (10) endemic species.

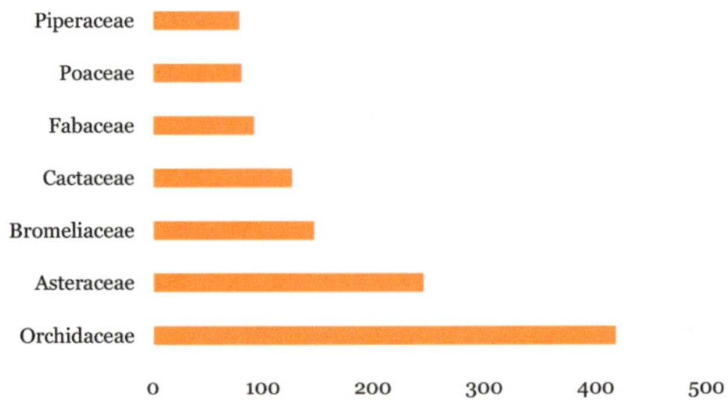

Figure 5.
The ten families with the largest number of endemic species in six vegetation formations of Bolivia.

	Endemic species
With four contiguous vegetation types	*Acianthera boliviana* (Rchb. F.) Pridgeon & M.W. Chase, *Aechmea kuntzeana* Mez, *Aegiphila herzogii* Moldenke, *Aristida friesii* Hack. Ex Henrard, *Bellucia beckii* S.S. Renner, *Cleistocactus samaipatanus* (Cárdenas) D.R. Hunt, *Croton abutilifolius* Croizat, *Furcraea boliviensis* Ravenna, *Gentianella inaequicalyx* (Gilg) J.S. Pringle, *Hippeastrum evansiae* (Traub & I.S. Nelson) H.E. Moore, *Lepidaploa tarijensis* (Griseb.) H. Rob., *Lonchocarpus pluvialis* Rusby, *Lupinus buchtienii* Rusby, *Maytenus tunarina* Loes. ex Kuntze, *Nassella holwayii* (Hitchc.) Barkworth, *Paranephelius asperifolius* (Muschl.) H. Rob. & Brettell, *Pitcairnia cardenasii* L.B. Sm., *Schoepfia tetramera* Herzog, *Steinbachiella leptoclada* Harms, *Stevia sarensis* B.L. Rob., *Tillandsia hegeri* Ehlers, *Trichogonia capitata* (Rusby) B.L. Rob.
With five contiguous vegetation types	*Festuca fiebrigii* Pilg., *Gentianella silenoides* (Gilg) Fabris, *Machaerium latifolium* Rusby, *Recordia boliviana* Moldenke, *Stevia setifera* Rusby ex B.L. Rob.
With six types	*Bougainvillea* modesta Heimerl, *Paspalum ekmanianum* Henrard

Table 2.
List of endemic species represented in less geographically restricted sites.

Yungueñan "páramo": Between 3500 and 4200 m elevation, this formation forms belts of grass and scrubs fed by nearly permanent precipitation. The most common genera include *Jarava, Festuca, Brachyotum, Clinopodium, Mutisia, Chuquiraga, Baccharis, Calceolaria*, and *Gnaphalium*. Other notable species include communities of *Puya raimondii* Harms and many species of the Ericaceae family.

Here a total of 846 species with ca. 22% endemics is found. The largest number of endemic species belongs to the family Orchidaceae (with 52 species), followed by Asteraceae (19) and Bromeliaceae (15) (**Figure 5d**). Among genera with the most numerous endemic species are *Puya* (Bromeliaceae) and *Gentianella* (Gentianaceae) with 15 and 13 species, respectively.

Tucumanian-Bolivian forest: Both humid to semi-humid forests cover the eastern slopes of the Andes in southern Bolivia between 500 and 1300 m elevation (**Figure 3d**). Among the more important elements are patches of *Polylepis hieronymi* Pilger (Rosaceae) and *Podocarpus parlatorei* Pilg. (Podocarpaceae) at higher altitudes; also are characteristic *Polylepis crista-galli* Ruiz & Pav., *Alnus acuminata* Kunth (Betulaceae) and *Juglans australis* Griseb. (Juglandaceae).

A total of 1647 species has been recorded here, of which close to 10% are endemic. Among the 130 families reported with endemic species, the most speciose are Orchidaceae (with 25), Asteraceae, and Bromeliaceae (21) (**Figure 5e**).

Begonia (Begoniaceae) and *Puya* (Bromeliaceae) are the genera with most endemic species (nine and eight, respectively).

Semideciduous Chiquitanian forest: It includes deciduous and semideciduous forests that are located throughout the Chiquitanian region in the department of Santa Cruz, between 400 and 700 m of elevation. Some characteristic species of the Chiquitania are *Machaerium nyctitans* (Vell.) Benth. (Fabaceae), *M. acutifolium* Vogel, *Amburana cearensis* (Allemão) A.C.Sm., *Schinopsis brasiliensis* Engel (Anacardiaceae), *Handroanthus lapacho* (K. Schumann) Sandwith (Bignoniaceae), and *Pseudobombax marginatum* (A. St.-Hil., A. Juss. & Cambess.) A. Robyns (Malvaceae).

Almost 6% are endemic, and the most common families with endemic species are Fabaceae (with 27 species), Asteraceae (16), and Malvaceae (14) (**Figure 5f**). Within the Fabaceae, the genus *Arachis* has the highest level of endemism with seven species.

3.1.1 Taxonomic groups

Among ferns and fern allies (Pteridophyta) together with the angiosperms of Bolivia, there are 2396 endemic species, in 670 genera and 141 families (**Table 3**). Seven angiosperm families account for 50% of the total (**Figure 6**): Orchidaceae, Asteraceae, Bromeliaceae, Cactaceae, Fabaceae, Poaceae, and Piperaceae. In the case of the Pteridophyta, two families comprise slightly less than 50% of the total: Dryopteridaceae with 43 species and Polypodiaceae with 26. Among the genera with the highest number of endemic species are *Puya* (Bromeliaceae, 55 species), *Lepanthes* (Orchidaceae, 52), *Peperomia* (Piperaceae, 51), *Solanum* (Solanaceae, 44), and *Tillandsia* (Bromeliaceae, 37), among others (**Figure 7**). In any case, the representation of four genera of the Orchidaceae has in total 134 endemic species.

If we relate the number of endemic species per family to the total numbers represented in Bolivia, the patterns show different percentages (**Table 4**). Orchids comprise the family with the highest number of endemic species, but this represents only 32.5% of the total; in Asteraceae the figure is 20.2%, whereas in the Bromeliaceae the endemic species amount to 45.8% of the total, and in the Cactaceae it rises to 55.9%. In Triuridaceae and Tropaeolaceae, the number of endemics amounts to about 50%, but the first has only 2 native species with 1 endemic, while the second has 14 species with 7 endemic. In addition there is another group where the percentage of endemism is low, ranging from 9 to 20% in families such as Fabaceae, Arecaceae, Poaceae, Amaranthaceae, and others.

In the case of species sorted by genera, various trends can be discerned (**Table 5**) as follows: 84.6% of *Puya* spp. (Bromeliaceae) are endemic, whereas in *Begonia* (Begoniaceae) the figure is 48.1%, *Piper* (Piperaceae) 30.1%, *Solanum* (Solanaceae) only the 23.7%, and *Trichilia* (Meliaceae) a mere 5.3%. Three genera, each with 25 endemic species, show very different trends in endemism, such as for *Masdevallia* (Orchidaceae) 78.1%, for *Siphocampylus* (Campanulaceae) 56.8%, and for *Miconia* (Melastomataceae) 17.7% (**Figure 8**).

	Families/genera/species	% from the total
Angiosperms	125/641/2263	89/96/94
Pteridophyta	16/29/139	11/4/6

Table 3.
Bolivian endemic vascular families, genera, and species and percentage from the total.

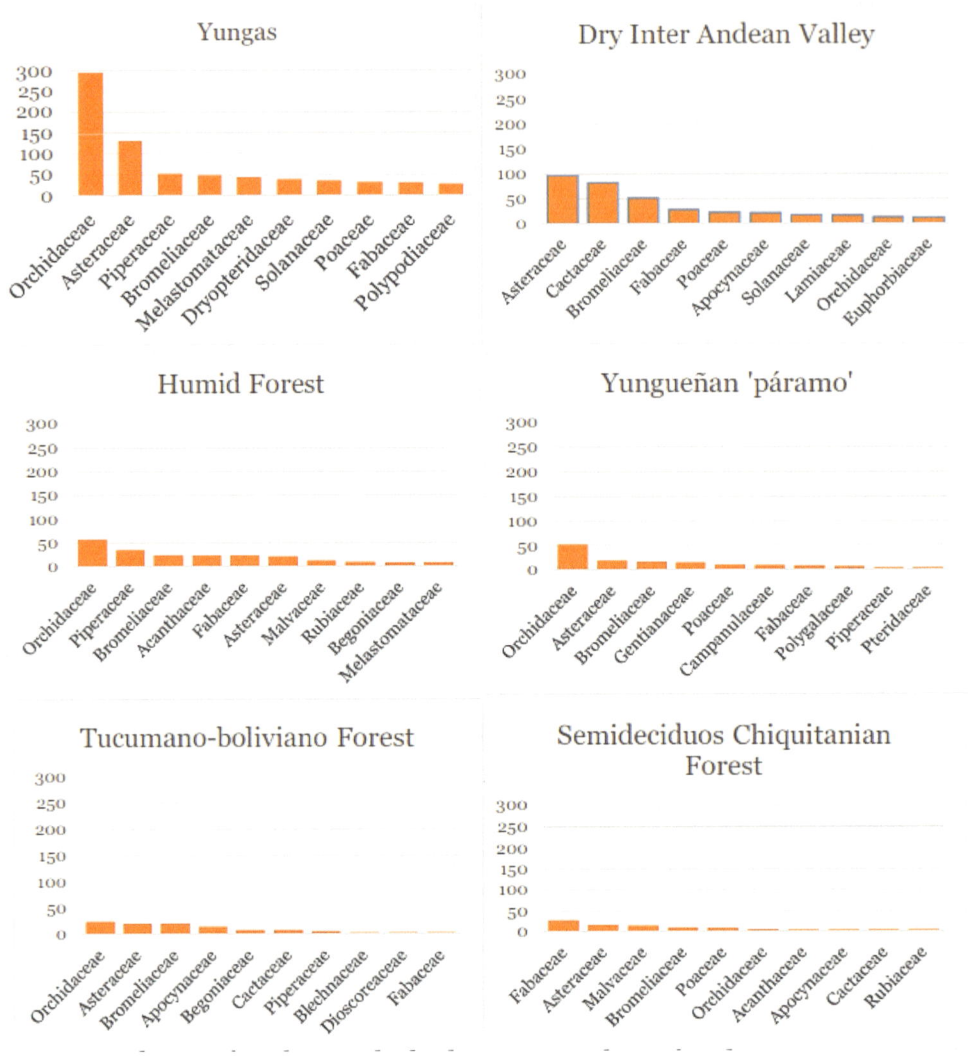

Figure 6.
Angiosperm plant families with the highest number of endemic species in Bolivia.

The records of endemic species of Bolivia are accompanied by a brief photographic synopsis that illustrates the characteristics of selected taxa (**Figures 9–12**). We present four groups of photographs: endemic palms [11], endemic plants of the Madidi region [10], endemic plants of the Bolivian Cerrado [16], and inter-Andean valleys with dry forests [23]. Tree palms and acaulescent species are represented here by *Attalea blepharopus* Mart. that grows in very humid forests in central Bolivia (**Figure 9a**) and then three species of the genus *Syagrus*: *S. petraea* (Mart.) Becc. from the plains and rocky hills of the Cerrado (**Figure 9b**), *S. yungasensis* M. Moraes from the drier montane forests (**Figure 9c**), and *S. cardenasii* Glassman from the alluvial plains, as well as in sub-Andean highlands (**Figure 9d**).

Among endemic Bolivian plants of the Madidi National Park in the NW of the country (mostly comprising Yungas vegetation), *Prestonia leco* A. Fuentes & J. F. Morales (Apocynaceae) is a liana found in the humid submontane forest (**Figure 10a**). *Passiflora madidiana* P. Jørg., Cayola & Araujo-Murak. (Passifloraceae) is an endemic climber of the dry forests of the Tuichi river basin (**Figure 10b**). *Stenostephanus suburceolatus* J.R.I.Wood (Acanthaceae) is a terrestrial

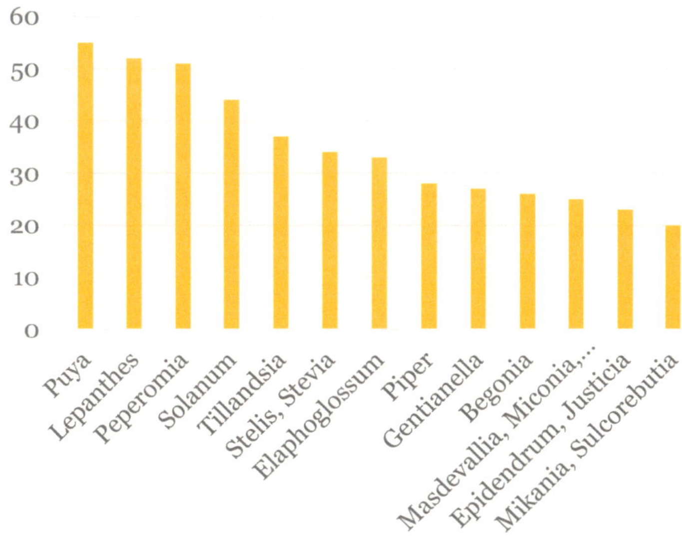

Figure 7.
Genera with the highest number of endemic species in Bolivia.

Plant family	Endemic species	Representativeness in the family (%)
Cactaceae	127	55.9
Triuridaceae, Tropaeolaceae	1, 7	50.0
Begoniaceae	26	48.1
Bromeliaceae	147	45.8
Zygophyllaceae	3	40.0
Proteaceae	5	38.5
Piperaceae	79	36.6
Orchidaceae	415	32.5
Acanthaceae	49	28.3
Dioscoreaceae	11	23.9
Asteraceae	243	20.2
Solanaceae	61	19.2
Euphorbiaceae	39	13.0
Amaranthaceae	14	11.4
Poaceae	81	9.5
Arecaceae	9	9.3
Fabaceae	92	8.9

Table 4.
Comparison of the level of endemism in some families of Bolivian native plants.

herb (**Figure 10c**), known from a single population of seasonally moist low altitude Andean forest. Finally, *Tristerix rhodanthus* Kuijt (Loranthaceae) is a hemiparasite collected in cloud forest fragments (**Figure 10d**).

The Chiquitanian endemic examples are as follows: *Blepharodon crabrorum* Goyder (Apocynaceae), a subshrub that blooms all the year and is found in rock

Plant genus	Endemic species	Representativeness in the genus (%)
Cleistocactus	14	82.3
Puya	55	84.6
Masdevallia	25	78.1
Lepanthes	52	74.3
Monnina	16	69.6
Arachis	12	63.1
Siphocampylus	25	56.8
Gentianella	27	50.0
Begonia	26	48.1
Peperomia	51	41.5
Tillandsia	37	35.2
Piper	28	30.1
Elaphoglossum	33	26.0
Solanum	44	23.7
Miconia	25	17.7
Pavonia	6	13.0
Trichilia	1	5.3

Table 5.
Comparison of the level endemism in some genera of Bolivian native plants.

crevices on vertical cliff faces (**Figure 11a**); *Mimosa crasspedisetosa* Fortunato & Palese (Fabaceae) a branched shrub or subshrub in scattered places in the plains of "campos cerrados" (**Figure 11b**); *Frailea chiquitana* Cárdenas (Cactaceae), a small and wooly cactus common in the slabs (**Figure 11c**); *Centratherum cardenasii* (I.S. Nelson & Traub) Van Scheepen (Amaryllidaceae), a perennial herb that grows in sandy Cerrado (**Figure 11d**); *Hippeastrum starkiorum* (I.S.Nelson & Traub) Van Scheepen (Amaryllidaceae), a rare bulbous herb that grows in rock crevices in campo rupestre and on hills (**Figure 11e**); and *Pitcairnia chiquitana* R. Vásquez & Ibisch (Bromeliaceae), cespitose plant, which is locally abundant on rock platforms and in campo rupestre (**Figure 11f**).

Endemic plants from dry forests in Andean valleys are as follows: *Cardenasiodendron brachypterum* (Loes.) F.A. Barkley (Anacardiaceae) (**Figure 12a**), a tree that grows on dry hillsides on rocky soils below 2600 m; *Mastigostyla cardenasii* R.C. Foster (Iridaceae) (**Figure 12b**), an herb of these valleys that reaches the humid puna; *Ipomoea exerta* Goyder & Fontella (Apocynaceae), a very rare herb and grows in dispersed zones at 2500 m (**Figure 12c**); *Oxypetalum fuscum* Epling (Lamiaceae), a short vine that is frequent in thickets (**Figure 12d**); *Lepechinia bella*, a small shrub that grows on rocky slopes (**Figure 12e**); and *Puya weddelliana* (Baker) Mez (Bromeliaceae) (**Figure 12f**), a rosette plant, which grows in groups on rocky slopes.

The most endangered species fall into three categories: critically endangered (CR) with 19 endemic species, followed by threatened (EN) with 80, and vulnerable (VU) with 66 (**Figure 12**). The total of 165 endemic plants that have been evaluated as threatened in Bolivia indicates that only 6.9% of the total number of endemic species has been assessed. A sample of 14 endemic species from Bolivia along with the category they belong to is shown in **Table 6**.

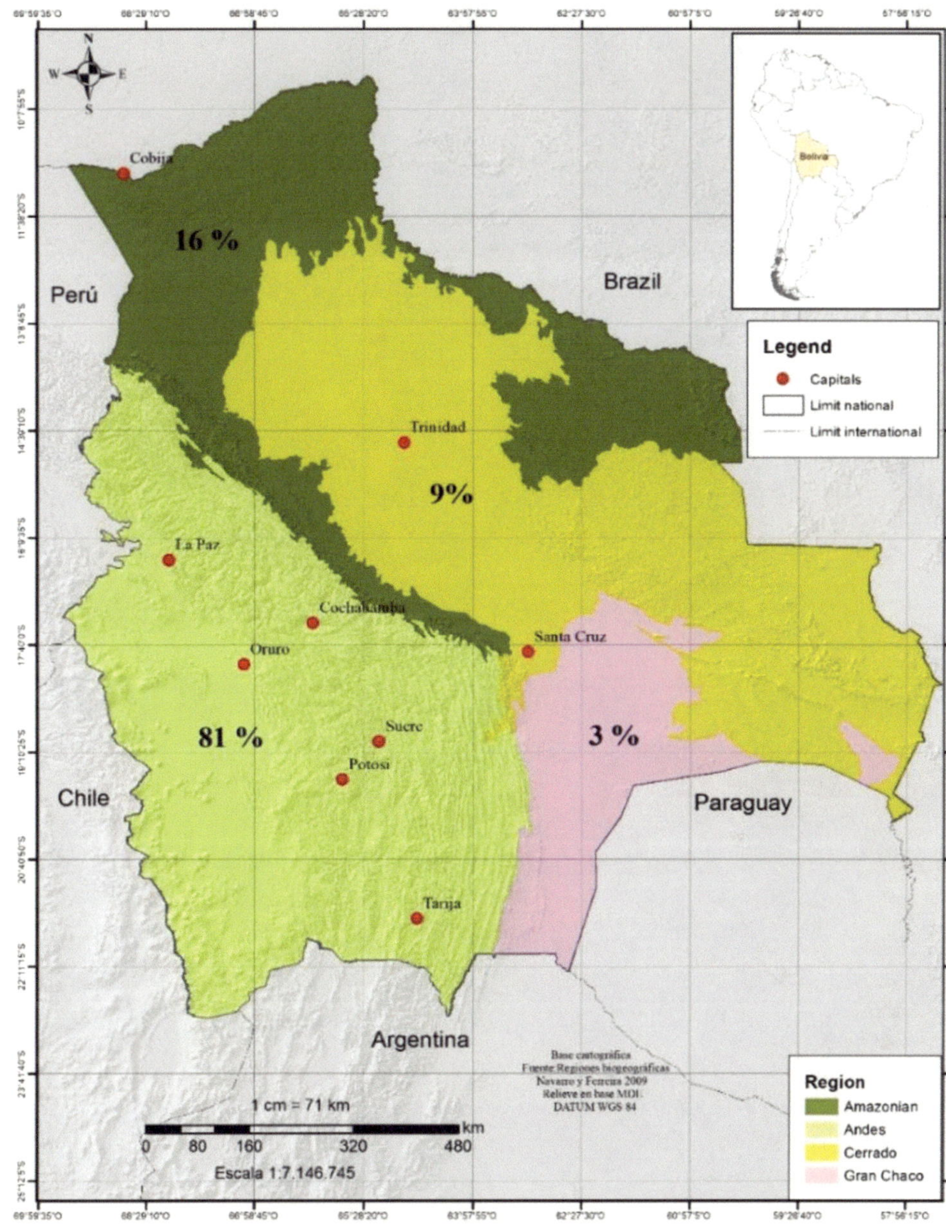

Figure 8.
Major biogeographic regions represented in Bolivia (adapted from [3]).

4. Discussion

Our knowledge of the floristic composition of Bolivia and its richness has increased in the last 20 years although there are still changes, such as new species and endemics; in addition, the level of our understanding has resulted from the intensification of botanical collections and fieldwork in geographical areas that are botanically less known. Until 1992, a list of 20 endemic monotypic genera of Bolivia in 13 families of vascular plants that existed was the first basis and a large number of herbarium specimens that had not been identified [1]; in the case of mosses, Bolivia has an

Figure 9.
Endemic palms of Bolivia. (a) Attalea blepharopus, *(b)* Syagrus petraea, *(c)* Syagrus yungasensis, *and (d)* Syagrus cardenasii *(Arecaceae). Photographs: Mónica Moraes R.*

inventory of 920 species, of which 55 are endemic [24]. In contrast in 2014, 2343 species of vascular plants were reported as endemic to Bolivia from a total of 12,165 species in 286 families [8], and 13.6% endemics were reported in the fern family Polypodiaceae and 33% for orchids but in the gymnosperms and in the angiosperms. Currently the number has risen to 12,239 native species of vascular plants (www.

Figure 10.
Endemic plants of Madidi National Park. a. Prestonia leco *(Apocynaceae), b.* Passiflora madidiana *(Passifloraceae), c.* Stenostephanus suburceolatus *(Acanthaceae) and d.* Tristerix rhodanthus *(Loranthaceae). Photographs, a,c-d: Alfredo Fuentes, b: Alejandro Araujo-Murakami.*

tropicos.org), and the present work reports a total of 2402 endemic plants of Bolivia; this means that in 4 years it has increased by 2.5%.

As in other botanical cases, it is very important to update the knowledge about the endemic flora. With the example of the 20 species in monotypic genera reported by Moraes and Beck [1], the number of endemisms for Bolivia was reduced to 14, and only 5 are monotypic: *Cardenasiodendron brachypterum* (Loes.) F.A. Barkley, *Polyclita turbinata* (Kuntze) A.C. Sm., *Rusbya taxifolia* Britton, *Boelckea beckii* Rossow, and *Recordia boliviana* Moldenke; only *Cardenasiodendron brachypterum* (Loes.) F.A. Barkley was assessed as VU by Navarro et al. [21].

On the eastern slope of the Andes and in the Amazon basin in Peru and Bolivia, 435 species in four plant groups, Anacardiaceae, Chrysobalanaceae, *Inga* (Fabaceae), and Malpighiaceae, were found to be endemic in the lowlands [25]. Acanthaceae presented its highest point of endemism at medium elevations, and nine plant groups, Aquifoliaceae, Brunelliaceae, Campanulaceae, Ericaceae, Loasaceae, Marcgraviaceae, *Fuchsia* (Onagraceae), and Passifloraceae, presented their highest point of endemism at elevations above 2000 m. Probably 20–25% of

Figure 11.
Endemic plants of the Chiquitano and Cerrado region. (a) Blepharodon crabrorum *(Apocynaceae), (b)* Mimosa crasspedisetosa *(Fabaceae), (c)* Frailea chiquitana *(Cactaceae), (d)* Centratherum cardenasii *(Asteraceae), (e)* Hippeastrum starkiorum *(Amaryllidaceae), and (f)* Pitcairnia chiquitana *(Bromeliaceae). Photographs: John Wood.*

the total vascular plants of Bolivia could be restricted to the country [26], the family with the highest number of endemic species being the Orchidaceae (35%, [27]). Plant endemism ranges between 59 and 85% of the species sampled in three areas of the Bolivian Chaco region [28], whereas only 18% of species in dry inter-Andean valleys of the western side of the country are endemic [29].

Our results support former conclusions by other authors. According to [26, 30], the highest concentration of plant endemism is registered in the Andean mountains, where both the Yungas (humid forests) and inter-Andean dry forests are found. Moreover, the major record of scientific collections and knowledge comes from the eastern slopes of the Andes from the NW toward the center of the country, where Bolivia's greatest biodiversity is higher [31]. Also the distribution of endemic palms

Figure 12.
Endemic plants of dry forests in inter-Andean valleys. (a) Cardenasiodendron brachypterum *(Anacardiaceae), (b)* Mastigostyla cardenasii *(Iridaceae), (c)* Ipomoea exerta *(Convolvulaceae), (d)* Oxypetalum fuscum *& Fontella (Apocynaceae), (e)* Lepechinia bella *(Lamiaceae), and (f)* Puya weddelliana *(Bromeliaceae). Photographs: John Wood.*

(Arecaceae) is associated with the eastern Andes [32]. But high-priority areas for representativeness of ecosystems and species, the Yungas forests, stand out as the main center of biological diversity in Bolivia due to the greater richness of species and better state of conservation with local centers of endemism, in addition to the Tucumano-Bolivian forest, Llanos de Moxos, and Amazonian forests with up to

Endemic species	Family	Category
Abarema centiflora Barneby & J.W. Grimes	Fabaceae	EN
Acanthosyris asipapote M. Nee	Santalaceae	CR
Arachis ipaensis Krapov. & W.C. Greg.	Fabaceae	EN
Begonia baumannii Lemoine	Begoniaceae	VU
Brunellia boliviana Britton ex Rusby	Brunelliaceae	VU
Nasa herzogii (Urb. & Gilg) Weigend	Loasaceae	EN
Passiflora chaparensis R. Vásquez	Passifloraceae	CR
Philibertia zongoensis Goyder	Apocynaceae	CR
Polylepis neglecta M. Kessler	Rosaceae	VU
Parajubaea sunkha M. Moraes	Arecaceae	EN
Roupala filiflora K.S. Edwards & Prance	Proteaceae	CR
Siphocampylus reflexus Rusby	Campanulaceae	EN
Trichocereus werdermannianus Backeb.	Cactaceae	VU
Zanthoxylum aculeatissimum Engl.	Rutaceae	EN

Table 6.
Endemic plants of Bolivia and UICN categories, based on [21, 22].

15% representativeness, while in the Altiplano, more fragmented and dispersed areas were identified [30].

Worldwide there is increasing evidence confirming that endemism is a powerful tool for use in global conservation efforts: hotspots based on levels of endemism cover more number of species than richness-based hotspots and are closely related to the degree of threat [33]. Therefore, a positive relationship between endemism and species richness might be expected [34]. Although Bolivia has not been completely surveyed floristically, 2403 endemic species have been recognized, and this may indicate that species richness may be greater than it has been documented. However, it is essential to establish the state of conservation of these species in order to increase efforts and generate more responsible actions to safeguard the natural heritage.

The confirmation on the presence of species and their status as endemics represents a greater effort by scientists and is an ongoing work. Each study and botanical survey contributes to the documentation of the flora of Bolivia. To develop an adequate conservation strategy, it will be necessary to assess this large residue to understand the threat level they face. However, there is very limited information and few collections of most endemic species. The humid montane forests are a hotspot in the tropical Andes that constitute a very rich region in endemism and would be supported by a greater total richness of plants. Therefore, it is essential to intensify the survey of species throughout their range, especially when there are still a large percentage of well-conserved landscapes and the threats are more locally concentrated, but the incidence of global warming against them with consequences can be catastrophic.

5. Conclusions

In terms of the concentration of the botanical collections made in Bolivia, there are large areas that still need to complete the surveys and records of species.

Therefore, it is expected that the total of endemic species will increase, especially in inter-Andean valleys with humid forests, as well as in the different formations that are represented in the east on Precambrian rocks of the Cerrado region in the country.

At the moment, the trends that have been interpreted according to the 2402 endemic plants are indicative with respect to the families and genera with the greatest number of species. 50% is represented by seven families; among them the Orchidaceae (418 spp.), Asteraceae (246 spp.), and Bromeliaceae (147) stand out. The representativeness on the total of native species is the following: Orchidaceae with 32.5%, Asteraceae 20.2%, and Bromeliaceae 45.8%, while among the genera, *Puya* (Bromeliaceae) presents 55 endemic species (representing 85% of the total native species of Bolivia), *Lepanthes* (Orchidaceae) with 52 (means 74%), and *Peperomia* (Piperaceae) with 51 (41.5%).

Although more than 80% of the plants come from the biogeographic region of the Andes, the landscape with the largest number of species (69%) is restricted to the eastern slopes (from 500 to 3500 m altitude), leaving the altiplanic landscape with only 7%. The geographical pattern that concentrates the 51% of endemisms in the humid montane forests of Yungas, where both Andean and Amazonian elements converge, is also fundamental. This feature of endemic plants also has relevance for the 22% that is represented in dry forests in Andean valleys, which corresponds to one-fifth of the total represented in Bolivia.

Finally, regarding the evaluation of the state of conservation of endemic plants in Bolivia, the efforts are still incipient. A total of 154 endemic species (6.9%) were evaluated in their conservation status according to IUCN categories, with the majority in threatened state (EN, 48.5%), followed by the vulnerable (VU, 40%) and in critical condition (CR, 11.5%).

Conflict of interest

The authors of this contribution indicate that they have no conflicts of interest that affect the content and information of this publication.

Author details

Mónica Moraes R.*, Carla Maldonado and Freddy S. Zenteno-Ruiz
Facultad de Ciencias Puras y Naturales, Herbario Nacional de Bolivia, Instituto de Ecología, Universidad Mayor de San Andrés, La Paz, Bolivia

*Address all correspondence to: mmoraes@fcpn.edu.bo

References

[1] Moraes RM, Beck S. Diversidad florística de Bolivia. In: Marconi M, editor. Conservación de la Diversidad Biológica en Bolivia. La Paz: CDC-Bolivia/USAID-Bolivia; 1992. pp. 73-111

[2] Navarro G. Vegetación y unidades biogeográficas. In: Navarro G, Maldonado M, editors. Geografía Ecológica de Bolivia: Vegetación y Ambientes Acuáticos. Cochabamba: Editorial Centro de Ecología Simón I. Patiño; 2002. pp. 1-500

[3] Navarro G, Ferreira W. Biogeografía de Bolivia. In: Moraes RM, Mostacedo B, Zapata Ferrufino B, Altamirano S, editors. Libro Rojo de Parientes Silvestres de Cultivos de Bolivia. La Paz: Ministerio de Medio ambiente y Agua. Vice ministerio de Medio Ambiente, Biodiversidad y Cambio Climático; 2009. pp. 23-39

[4] Isacks BL. Uplift of the central Andean plateau and bending of the Bolivian orocline. Journal of Geophysical Research. 1988;**93**: 3211-3231

[5] Hazzi NA, Moreno JS, Movliav CO, Palacio RD. Biogeographic regions and events of isolation and diversification of the endemic biota of the tropical Andes. PNAS. 2018;**115**(31):7985-7990. DOI: 10.1073/pnas.1803908115

[6] Wood JRI, Mamani F, Pozo P, Soto D, Villarroel D. Guía Darwin de las Plantas de los Cerrados de la Chiquitania. Santa Cruz: Editorial Museo de Historia Natural Noel Kempff Mercado; 2011

[7] Pozo P, Wood J, Soto D, Beck S. Plantas endémicas de afloramientos rocosos en las serranías de roboré y concepción: Implicaciones para su conservación. Revista de la Sociedad Boliviana de Botánica. 2013;7(1):73-81

[8] Jørgensen PM, Nee M, Beck SG. Resultados. In: Jørgensen PM, Nee MH, Beck SG, editors. Catálogo de las Plantas Vasculares de Bolivia. St. Louis: Monographs in Systematic Botany from the Missouri Botanical Garden; 2014. pp. 33-76

[9] Atahuachi M, Van der Bent ML, Wood JRI, Lewis GP, Hughes CE. Bolivian *Mimosa* (Leguminosae, Mimosoideae): Three new species and a species checklist. Phytotaxa. 2016; **260**(3):201-222. DOI: 10.11646/phytotaxa.260.3.1

[10] Fuentes AF. Flora y vegetación leñosa de los bosques de los Andes en la región Madidi, La Paz (Bolivia) [thesis]. Madrid: Facultad de Farmacia, Departamento de Biología Vegetal, Universidad Complutense de Madrid; 2016

[11] Moraes RM. Actualización de la lista de especies de Arecaceae para Bolivia. Revista de la Sociedad Boliviana de Botánica. 2015;**5**(1):19-28

[12] Moraes RM, Pintaud JC. *Attalea blepharopus* Mart. (Arecaceae) from Bolivia revisited since Martius. Candollea. 2016;**71**:27-32

[13] Noblick LR. A revision of the genus *Syagrus* (Arecaceae). Phytotaxa. 2017; **294**:1-262

[14] Santamaría-Aguilar D, Fuentes AF, Lagomarsino LP. Three new species of *Freziera* (Pentaphylacaceae, Freziereae) from Bolivia and Peru. Phytotaxa. 2018; **39**(2). DOI: 10.11646/phytotaxa.349.2.1

[15] Villarroel D, Gomes-Bezerra KM. New botanical discoveries of Myrtaceae from Bolivia and notes on *Psidium hians*. Phytotaxa. 2015;**195**(2):163-170

[16] JRI W, editor. Guía Darwin de las Plantas de los Cerrados de la

Chiquitania. Santa Cruz: Museo de Historia Natural Noel Kempff Mercado, Universidad Autónoma Gabriel René Moreno – Department of Plant Sciences, Oxford University; 2011. p. 212

[17] Wood JRI, Carine MA, Harris D, Wilkin P, Williams B, Scotland RW. *Ipomoea* (Convolvulaceae) in Bolivia. Kew Bulletin. 2015;**70**(31):1-123

[18] Ibisch PL, Beck SG, Gerkmann B, Carretero A. Ecoregiones y ecosistemas. In: Ibisch P, Mérida G, editors. Biodiversidad: La Riqueza de Bolivia. Estado de Conocimiento y Conservación. Santa Cruz: Fundación Amigos de la Naturaleza; 2003. pp. 47-88

[19] Mamani F, Pozo P, Soto D, Villarroel D, Wood JRI, editors. Libro Rojo de las Plantas de los Cerrados del Oriente Boliviano. Santa Cruz: Museo de Historia Natural Noel Kempff Mercado; 2010. p. 212

[20] Moraes RM, Mostacedo B, Zapata Ferrufino B, Altamirano S. Libro Rojo de Parientes Silvestres de Cultivos de Bolivia. La Paz: Ministerio de Medio ambiente y Agua. Vice Ministerio de Medio Ambiente, Biodiversidad y Cambio Climático; 2009

[21] Navarro G, Arrázola S, Atahuachi M, De la Barra N, Mercado M, Ferreira W, et al. Libro Rojo de la Flora Amenazada de Bolivia. Volumen I—Zona andina. Ministerio de Medio Ambiente y Agua—Rumbol Srl: La Paz; 2012. p. 584

[22] Centro de Biodiversidad y Genética. Libro Rojo de la Flora Amenazada de Bolivia. Volumen II—Tierras Bajas. Cochabamba; 2018, 2018. p. 708

[23] Wood JRI, editor. La Guía Darwin de las Flores de los Valles Bolivianos. London: University of Oxford and Darwin Initiative; 2005. p. 187

[24] Aldana C, Calzadilla M, Churchill SP. Evaluación de los musgos endémicos de Bolivia. Revista de la Sociedad Boliviana de Botánica. 2011;**5**(1):53-67

[25] Natureserve. Especies Endémicas y Sistemas Ecológicos en la Vertiente Oriental de Los Andes y la Cuenca del Amazonas en Perú y Bolivia. Arlington, Virginia; 2007

[26] Ibisch PL, Beck SG. Espermatófitas. In: Ibisch P, Mérida G, editors. Biodiversidad: La Riqueza de Bolivia. Estado de Conocimiento y Conservación. Santa Cruz: Fundación Amigos de la Naturaleza; 2003. pp. 103-112

[27] Ibisch PL. Erhaltung der pflanzlichen vielfalt des megadiversitätslandes bolivien. problemanalyse und bewertungsmethoden sowie erhaltungsstrategien und ökoregionale leitbilder [thesis]. Bonn: Faculty of Math and Natural Sciences, University of Bonn; 2002

[28] Parker TA, Gentry AH, Foster RB, Emmons LH, Remsen JV. The lowland dry forests of Bolivia: A global conservation priority. In: Conservation International, RAP Working Papers 4; 1993

[29] López RP. Diversidad florística y endemismo de los valles secos bolivianos. Ecología en Bolivia. 2003; **38**(1):27-60

[30] Araujo N, Müller R, Nowicki C, Ibisch P, editors. Prioridades de Conservación de la Biodiversidad en Bolivia. Cuidando la Madre Tierra. Santa Cruz: Editorial Fundación Amigos de la Naturaleza; 2010. p. 74

[31] Fernández M, Navarro LM, Apaza-Quevedo A, Gallegos SC, Marques A, Zambrana-Torrelio C, et al. Challenges and opportunities for the Bolivian biodiversity observation network.

Journal of Life on Earth Biodiversity. 2015;**16**. DOI: 10.1080/14888386.2015.1068710

[32] Moraes RM, Rios-Uzeda B, Moreno LR, Huanca-Huarachi G, Larrea-Alcázar D. Using potential distribution models for patterns of species richness, endemism, and phytogeography of palm species in Bolivia. Tropical Conservation Science Journal. 2014;7(1):45-60

[33] Orme CDL, Davies RG, Burgess M, Eigenbrod F, Pickup N, Olson VA, et al. Global hotspots of species richness are not congruent with endemism or threat. Nature. 2005;**436**:1016-1019. DOI: 10.1038/nature03850

[34] Lamoreux JF, Morrison JC, Ricketts TH, Olson DM, Dinerstein E, McKnight MW, et al. Global test of biodiversity concordance and the importance of endemism. Nature. 2006;**440**:212-214. DOI: 10.1038/nature04291

Chapter 4

The Endemism of the Vascular Flora of Mexico Present in Comarca Lagunera, an Agricultural Region in the Chihuahuan Desert

Alberto González-Zamora and Rebeca Pérez-Morales

Abstract

A study of the distribution of 321 taxa of endemic vascular plants of Mexico distributed in Comarca Lagunera, a region of northern central Mexico within the Chihuahuan Desert, was conducted. The analysis consisted in detecting the areas of high richness and with this information propose areas for the conservation of plant biodiversity in this region. The study includes an analysis of species richness at the level of political units (municipalities), vegetation types, and grid cells of 10 × 10 km. Additionally, the corrected weighted endemism index was calculated using the grid cells. The sites with the richest taxa are located in the mountain areas; however, these do not coincide with the sites with the highest index of endemism since a high percentage of taxa have a restricted distribution to one of the proposed units. Thirty-six taxa are recognized with restricted distribution to the boundaries of Comarca Lagunera, most of them considered as microendemics, which have been described in recent years. Therefore, it is necessary to establish biodiversity conservation programs in the region since much of Comarca Lagunera territory is dedicated to agricultural and industrial activities.

Keywords: biodiversity informatics, conservation, Coahuila, corrected weight endemism index, Durango, north of Mexico, chorological analysis, semiarid vegetation

1. Introduction

Biodiversity informatics is a set of tools that allows free and rapid access to knowledge accumulated in different sources and media such as biological collections and specialized literature, which has facilitated the arrangement, management, analysis, and interpretation of biodiversity in addition to the generation of models; many of them focused on the conservation of biodiversity [1, 2]. The increase in knowledge of biodiversity is seen in a large amount of information from different databases available on the Internet, of which the Global Biodiversity Information Facility (GBIF–www.gbif.org) stands out, housing more than 1000 million records of species distribution (October 2018). This number of records continues growing, as it improves the quality control of the data hosted because more and more institutions deposit their data on that platform. Another tool that is very useful in biodiversity informatics is the Biodiversity Heritage Library (BHL—www.biodiversitylibrary.org), which allows free access to millions of published historical documents related to taxonomy, biogeography, and ecological aspects of the groups that form the biodiversity, among other items; thanks to these platforms, it is now easier to consult information that in the past was difficult to access for many research groups.

In Mexico, the National Commission for Knowledge and Use of Biodiversity (CONABIO) carried out one of the first attempts to gather and make available information to researchers about the distribution of Mexican species. Currently, the National System of Information on Biodiversity (SNIB) is one of the most reliable databases that houses information on the distribution of a large number of species of virtually all biological groups known to Mexico; this database gathers information from a large number of Mexican collections in which the taxonomic information has been reviewed, identified, and determined by specialists, and it also has a high level of quality in the revision phase of the georeferencing of the localities of distribution, so its reliability is very high [3]. The effort made by different institutions to learn about biodiversity allows Mexico to contribute to one of the objectives of the Systematic Agenda [4].

Although Mexico has a great botanical tradition and the knowledge of the flora in several regions is relatively well known, such as the north of the country and mainly the Chihuahuan Desert [5, 6], there are areas that do not still have an inventory of their flora, and therefore the distribution of the endemic elements is unknown, which is an impediment to carry out conservation plans. One of the regions in which a complete inventory of the flora is not yet made is Comarca Lagunera, a region located in the northern center of Mexico, within the Chihuahuan Desert (**Figure 1**).

The boundaries of Comarca Lagunera differ depending on the approach considered, and in this chapter, the limits proposed by Sánchez [7] are taken, which recognize that Comarca Lagunera is composed of 15 municipalities that share particular natural and socioeconomic attributes. Five of these municipalities belong to the state of Coahuila (Francisco I. Madero, Matamoros, San Pedro, Torreón, and Viesca) and

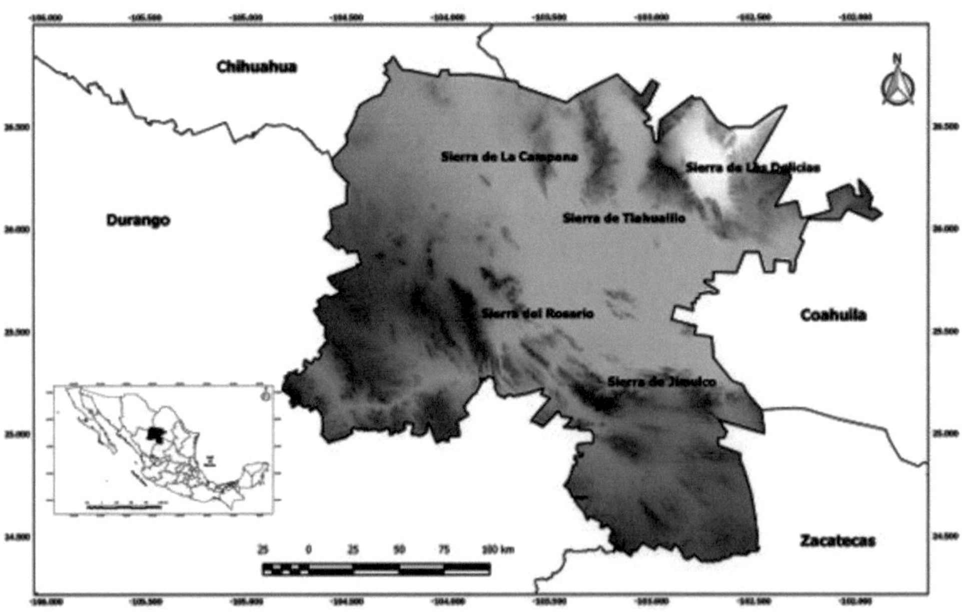

Figure 1.
Localization of Comarca Lagunera in México and its principal mountain zones.

10 to the state of Durango (General Simón Bolívar, Gómez Palacio, Lerdo, Mapimí, Nazas, Rodeo, San Juan de Guadalupe, San Luis del Cordero, San Pedro del Gallo, and Tlahualilo) (**Figure 2**). Under these limits, Comarca Lagunera covers an area of just over 43,000 km^2, which is equivalent to the total territory of Denmark.

The climate of Comarca Lagunera is semiarid to very arid, according to García [8] (**Figure 3**), with an average annual temperature of 20–22°C and an annual rainfall of 200–300 mm. The climate types are BS1hw (semiarid-semiwarm), BS1kw (semiarid-tempered), BS0hw (arid-semiwarm), BS0kw (arid-tempered), and BWhw (arid-semiwarm). The relief of the region is composed of small hills and low elevations, as well as adjacent depressions [9], among the main elevations are the mountains of Candelaria, El Rosario, El Sarnoso, España, Jimulco, La Campana, Las Cadenas, Las Delicias, Mapimí, and Tlahualilo.

Mexico has a richness flora and vegetation as a result of the great diversity of ecological conditions, which in turn has been shaped by the complex geological history of its territory. The different types of vegetation that are recognized are based on ecological, floristic, and physiognomic affinities; although some consider the edaphological, geological, and topographic aspects of greater importance [10]. The different vegetation types according to the National Institute of Statistic and Geography (INEGI) are based on [11, 12]. The land use and vegetation information elaborated by the INEGI is a reliable element in terms of information and its feasibility; therefore, they are used as a basic framework by government agencies, the private sector, and academia, and it is a basic input for the implementation of institutional programs as a reference for the preparation of reports on the state of plant cover in Mexico [13]. The diversity of types of vegetation in Mexico is so wide that it is difficult to find an equivalence with those that exist in other places [14], including those of North America in northern Mexico [15].

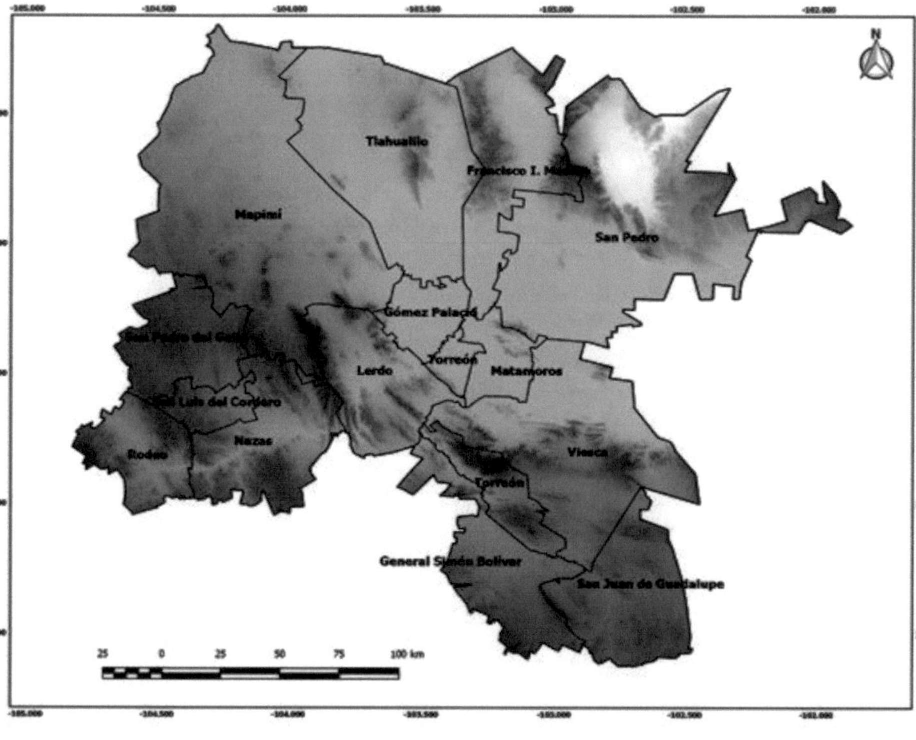

Figure 2.
Municipalities that form Comarca Lagunera, according to Sánchez [7].

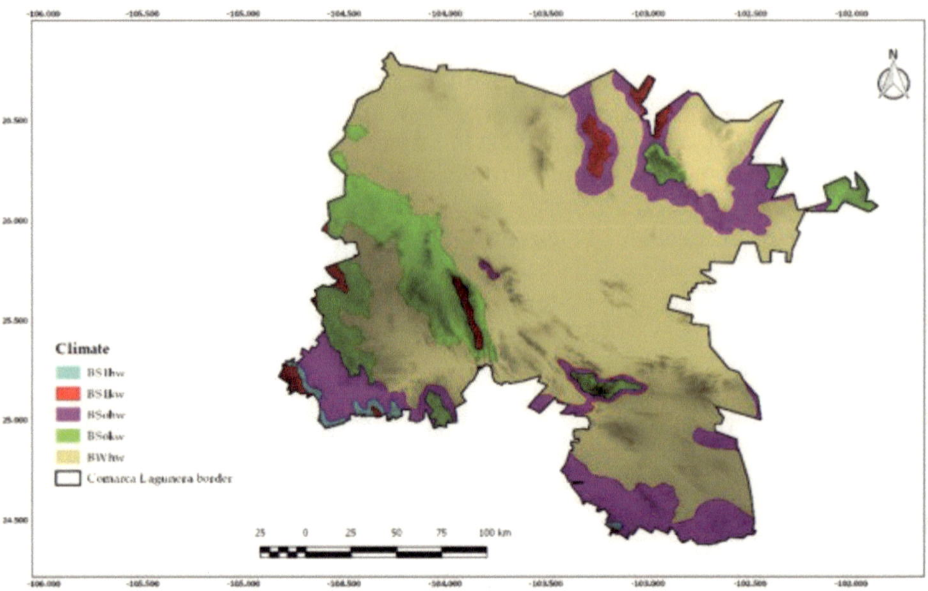

Figure 3.
Climates in Comarca Lagunera according to García [8].

According to the National Institute of Statistic and Geography (INEGI) [13], the land use and vegetation types in the area are 17. The description of the types of vegetation exceeds the objective of this work so only listed below, in parentheses are the most important genera that are present in each of them: chaparral (*Arctostaphylos* Adans., *Ceanothus* L., *Cercocarpus* Kunth, *Eriogonum* Michx., *Heteromeles* M. Roem., *Mimulus* L., *Quercus* L., *Rhamnus* L., and *Yucca* L.), crasicaule scrub (*Acacia* Mill., *Brickellia* Elliot, *Buddleja* L., *Celtis* L., *Dalea* Juss., *Larrea* Cav., *Mimosa* L., *Opuntia* Mill., *Prosopis* Burkart, *Rhus* L., *Stenocereus* (A. Berger) Riccob., and *Yucca*), gallery vegetation (*Acacia*, *Chilopsis* Don, *Mimosa*, *Prosopis*, *Salix* L., and *Senecio* L.), gypsophile vegetation (*Dicranocarpus* A. Gray, *Flaveria* Juss., *Helianthemum* Mill., *Petalonyx* A. Gray, *Sartwellia* A. Gray, *Selinocarpus* A. Gray, and *Stipa* L.), halophile grassland (*Distichlis* Raf., *Hilaria* Kunth, *Spartina* Schreb. ex Gmel., *Sporobolus* R.Br., and *Uniola* L.), halophile xerophile vegetation (*Atriplex* L., *Batis* L., *Frankenia* L., *Spartina*, and *Zostera* L.), induced grassland (*Andropogon* L., *Aristida* L., *Bouteloua* Lag., *Bromus* L., *Buchloe* Engelm., *Calamagrostis* Adans., *Cathestecum* C. Presl, *Deschampsia* P. Beauv., *Erioneuron* Nash, *Festuca* L., *Hilaria*, *Lycurus* Kunth, *Muhlenbergia* Schreb. ex Gmel., *Stipa*, *Trachypogon* Nees, and *Trisetum* Pers.), microphile desertic scrub (*Acacia*, *Cercidium* Tul., *Condalia* Cav., *Chilopsis*, *Flourensia* DC., *Fouquieria* Kunth, *Hymenoclea* Torr. & A. Gray, *Larrea*, *Lycium* L., *Olneya* A. Gray, *Opuntia*, and *Prosopis*), mesquite forest (*Prosopis*), natural grassland (*Bouteloua*), oak forest (*Quercus*), oak-pine forest (*Quercus*, *Pinus* L.), pine forest (*Pinus*), rosetophile desertic scrub (*Agave* L., *Dasylirion* Zucc., *Euphorbia* L., *Hechtia* Klotzsch, *Parthenium* L., and *Yucca*), submontane scrub (*Acacia*, *Agave*, *Aristida*, *Bouteloua*, *Capparis* L., *Cordia* L., *Euphorbia*, *Flourensia*, *Gochnatia* Kunth, *Karwinskia* Zucc., *Leucophyllum* Humb. & Bonpl., *Mimosa*, *Mortonia* A. Gray, *Neopringlea S. Watson*, *Pithecellobium* Mart., *Rhus*, *Tridens* Roem. & Schult, and *Zanthoxylum* L.), vegetation of sandy desert (*Ambrosia* L., *Atriplex*, *Ephedra* L., *Larrea*, *Opuntia*, *Prosopis*, and *Yucca*), and xerophile scrub (*Acacia*, *Agave*, *Dasylirion*, *Euphorbia*, *Flourensia*, *Fouquieria*, *Larrea*, *Opuntia*, *Prosopis*, and *Yucca*). A considerable part of the territory of the region is occupied for agricultural and livestock activities (**Figure 4**).

Within the territory of Comarca Lagunera, five areas have been decreed for the protection of biodiversity: a protected natural area at the federal level (Reserva de la Biósfera Mapimí), one at the state level (Parque Estatal Cañón de Fernández), one

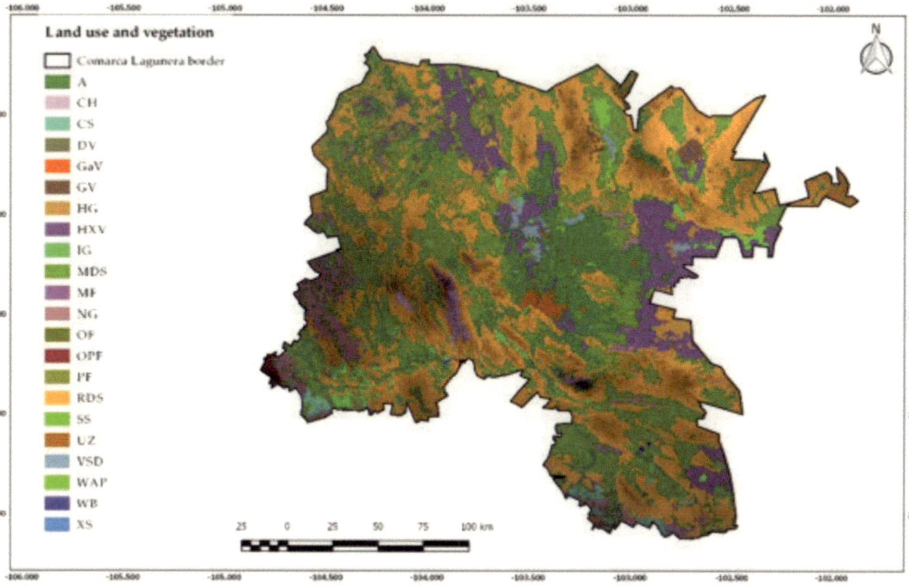

Figure 4.
Land use and vegetation types in Comarca Lagunera according to INEGI [10]. A = agricultural; CH = chaparral; CS = crasicaule scrub; DV = depleted of vegetation; GaV = gallery vegetation; GV = gypsophile vegetation; HG = halophile grassland; HXV = halophile xerophile vegetation; IG = induced grassland; MDS = microphile desertic scrub; MF = Mesquite forest; NG = natural grassland; OF = Oak forest; OPF = Oak-pine forest; PF = Pine forest; RDS = rosetophile desertic scrub; SS = submontane scrub; UZ = urban zones; VSD = vegetation of sandy desert; WAP = without apparent vegetation; WB = water body; and XS = xerophile scrub.

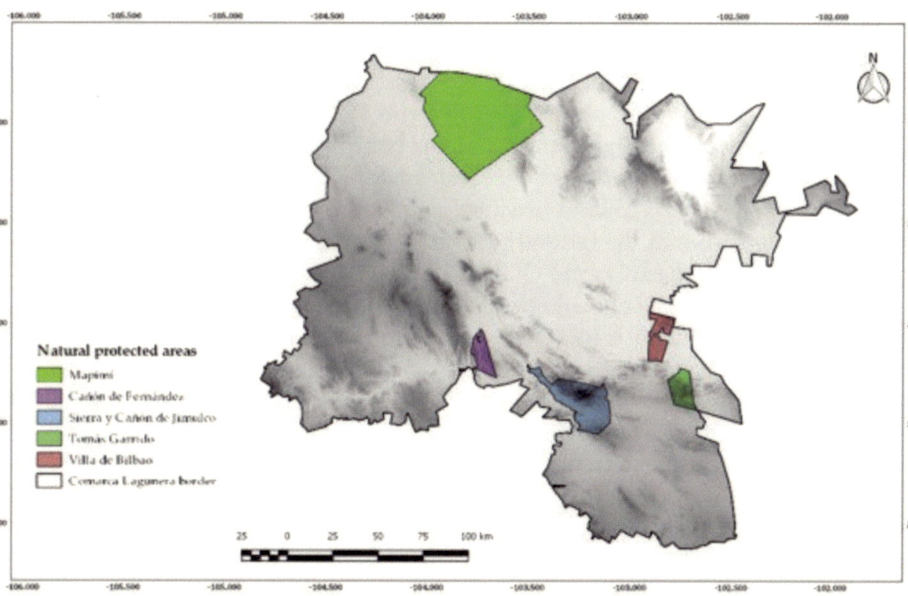

Figure 5.
Protected natural areas decreed in Comarca Lagunera.

municipal (Sierra de Jimulco), and two at ejidal level (Dunas de Bilbao and Ejido Tomás Garrido) (**Figure 5**).

The objective of the present study is to analyze the distribution and abundance of the taxa of the endemic flora of Mexico distributed in Comarca Lagunera in order to recognize potential sites for their conservation.

2. Methods

The search for information regarding the distribution of vascular plants in the SNIB database was carried out, where the records of the collected specimens in Comarca Lagunera were compiled. Considering the political limits of the 15 municipalities that form this region of Mexico, an exhaustive literature search was conducted in the BHL to detect nonregistered taxa, as well as verify the distribution of the taxa and addition of sites of collection not reported in the SNIB databases.

The information of the protologue of most of the taxa was reviewed, mainly those that have been described as new in the last decades, as well as the sites of register of collection of these taxa. The information published in various inventories at the state or regional level for both the flora as a whole or for specific groups was also consulted [6, 16–26], in addition to the review of the specimens deposited in the Interdisciplinary Research Center for Regional Integral Development (CIIDIR) herbarium.

Based on the information collected, a database with the records of the different taxa distributed in the boundaries of Comarca Lagunera was constructed. The nomenclatural information to avoid synonyms was based on the consultation of databases on the Internet of the International Plant Name Index (IPNI) (https://www.ipni.org), The Plant List (http://www.theplantlist.org), Tropicos (http://www.tropicos.org), and mainly in Villaseñor [27], which is so far the work that brings together the most extensive knowledge about the distribution of the flora of Mexico, where the author mentions that there are about 23,314 species and 1,414 subspecific categories.

In this study, the classification used by Villaseñor [27] for higher taxonomic categories is followed, which is based in turn mainly on the classification of Angiosperm Phylogeny Group [28]. The constructed database was refined to identify duplicate records and errors in the georeferencing. The records without geographic reference or with errors in it were georeferenced with the help of the topographic charts of INEGI scale 1:250,000 and 1:50,000 (www.inegi.org.mx). For maps elaboration, the digital model of the continuum of Mexican elevations (http://www.beta.inegi.org.mx/app/geo2/elevacionesmex) was used, in addition to the cartography related to land use and vegetation types for the region [13] and natural protected areas at the federal [29], state, municipal, and ejidal level [30]. All maps were created in the software QGIS 2.18.23 [31].

In order to perform the chorological analysis, the following were used as Operational Geographical Units (OGUs): the municipalities that are part of Comarca Lagunera, vegetation types, and a grid cell of 10 × 10 km latitude/longitude. Species richness was measured as the total count of species within each of the proposed units. In the analyzes at the municipal level, the criteria suggested by Dávila-Aranda et al. [32], with modifications due to the number of municipalities of Comarca Lagunera, to classify the level of restriction in the distribution of the species in scarcely distributed (taxa registered only in one municipality), closely distributed (2–4 municipalities), normally distributed (5–7 municipalities), and widely distributed (8 or more municipalities) were employed. Using these study units, a species accumulation curve was constructed, and with the aid of the nonparametric Chao2 estimator, the approximate number of endemic taxa, which are still unknown in Comarca Lagunera, was determined.

The vegetation types recognized by INEGI [13] for Comarca Lagunera were used; in each type of vegetation, the contribution of the endemic species of Mexico was evaluated. The information of the presence records was used to perform a cluster analysis, for which an absence-presence matrix was constructed to carry out a similarity analysis applying the Jaccard index, which was preferred over other indexes due to the fact that it does not consider shared absences. Once the similarity matrix was obtained, a grouping analysis (UPGMA) was carried out, to determine the floristic similarity among vegetation types. The corresponding phenogram was generated with the average linkage or UPGMA using the SAHN-clustering command in the NTSYSpc version 2.0 program [33].

Since the municipality size of Comarca Lagunera is not uniform and the area occupied by the different vegetation types is extremely different, it was necessary to carry out an alternative chorological analysis with units of standardized size. For this reason, a grid cell of 10 km per side was constructed to reduce the effect of the differences in the size of the units used for the analysis according to Crisp et al. [34]. An analysis of richness and endemism was also carried out following the proposal of Crisp et al. [34] and Linder [35], and this analysis called corrected weighted endemism index (CWEI) was applied for the first time to analyze the endemic elements of the flora of Australia and later in Africa. In Mexico, it has been applied in the analysis of diverse groups of flora and fauna, mainly to detect species-rich sites and a high level of endemism in order to propose sites for the conservation of biodiversity [36–39].

The CWEI [34, 35] is a methodology that calculates the relationship of taxa richness with weighted endemism. For this index, the presence of the taxon in an OGU is considered, and the abundance is omitted. It consists of three basic steps: in the first one, the index of endemism is calculated for each one of the taxa which is the result of the amplitude of its distribution, so it is calculated as the reciprocal of its distribution, and therefore, a species with restricted distribution to a single OGU (in this case grid cells) is assigned an index of "1," a species that is distributed in two OGUs is assigned an index of "0.5," and so on in such a way that the index will be lower as its distribution increases, so it is weighted with a higher value for species with restricted distribution; the second step is to calculate the weighted endemic index (WEI) for each of the OGUs, this is achieved by summing the values of the endemism index of each of the taxa that are distributed in each OGU; while the third step consists in dividing the WEI by each one of the OGUs among the richness of species that are distributed in it, in this way, we obtain the CWEI for each one of the OGUs; OGUs with the highest scores are recognized as centers of endemism.

Finally, a complementarity analysis was carried out [40] to detect the minimum number of grid cells needed to conserve most of the endemic species that inhabit this area of Chihuahuan Desert. The method consists in selecting the areas to be protected based on the total species richness from the following procedure: the grid cell with the highest number of taxa is selected, this is the grid cell with priority 1, the species registered in it are eliminated from the analysis, and then the selection of the grid cell with the highest number of remaining taxa (complement) that have not yet been selected in the previous step is repeated; if two or more grid cells have the same number of taxa, the first grid cell identified is selected, and the procedure ends when all the taxa have been selected.

3. Results and discussion

The richness of endemic plants of Mexico represented in Comarca Lagunera consists of 59 families, 184 genera, 300 species, and 21 infraspecific taxa (8 subspecies and 13 varieties) totaling 321 taxa (**Table 1**). Thirty-six taxa possess known distribution only for the area corresponding to Comarca Lagunera (Appendix), 21 more taxa are known only from the Chihuahuan Desert (*sensu* Villarreal-Quintanilla et al. [41]), 33 more taxa only occur in the states of Coahuila and Durango, and the rest of the taxa is restricted to the territory of Mexico but with a greatly wide distribution. This diversity of endemic taxa corresponds to approximately 10% of the total endemic species of Mexico [27] and to about 30% of the flora known for Comarca Lagunera [González-Zamora & Pérez-Morales, *in prep*.].

The families with the highest number of endemic taxa in Mexico distributed in Comarca Lagunera are Asteraceae and Cactaceae, which collectively account for approximately 35% of the total, this percentage is consistent with the results of Villareal-Quintanilla et al. [41] for the endemic flora of the Chihuahuan Desert, as well as for other semi-desert regions of central Mexico. The first six families with the highest number of endemic taxa represent about 54% of those reported for this region of Mexico and 36% of the taxa restricted to Comarca Lagunera (**Table 2**).

Of the 10 genera with 5 or more endemic species of Mexico, 4 correspond to the family Cactaceae, 2 to Asteraceae, 1 to Lamiaceae, 1 to Asparagaceae, 1 to Caryophyllaceae, and 1 to Fabaceae (**Table 3**). *Corynopuntia* F.M. Knuth and *Gaillardia* Foug. are the best represented genera; since in the study area, five of the seven species and five of the six species reported as endemic to Mexico, respectively, are distributed, including *Corynopuntia halophila* D. Donati which until now has only been reported as restricted to the territory of Comarca Lagunera [42], the same case is presented for *Salvia jessicae* B.L. Turner [43] and *Drymaria jenniferae* Villarreal & A.E. Estrada [44].

In the distribution by political entities, it was found that in general terms, the municipalities corresponding to the state of Coahuila have greater richness

Taxonomic group	Families	Genera	Species	Infraspecific taxa
Ferns	2	3	3	
Gymnosperms	1	1	1	
Angiosperms	56	180	296	21
Monocots	6	18	30	1
Eudicots	50	162	266	20
Total	**59**	**184**	**300**	**21**

Table 1.
Number of taxa of endemic vascular plants of Mexico in Comarca Lagunera.

Family	Taxa	Percentage with respect to the total	Genera represented	Endemic taxa of Comarca Lagunera
Asteraceae	69	21.5	38	6
Cactaceae	43	13.4	15	2
Fabaceae	27	8.4	16	3
Asparagaceae	13	4.0	7	1
Poaceae	10	3.1	6	
Lamiaceae	10	3.1	2	1
Total	**172**	**53.5**	**84**	**13**

Table 2.
Families with the highest number of endemic taxa of Mexico in Comarca Lagunera.

Genus (Family)	Species in Mexico	Species endemics of Mexico	Endemic taxa of Mexico in Comarca Lagunera
Coryphantha (Engelm.) Lem. (Cactaceae)	46	37	8
Mammillaria Torr. & A. Gray (Cactaceae)	169	145	8
Salvia L. (Lamiaceae)	328	258	8
Agave L. (Asparagaceae)	159	133	5
Brickellia Elliott (Asteraceae)	85	57	5
Gaillardia Foug. (Asteraceae)	12	6	5
Corynopuntia F.M. Knuth (Cactaceae)	10	7	5
Opuntia Mill. (Cactaceae)	91	62	5
Drymaria Willd. ex Schult. (Caryophyllaceae)	38	23	5
Dalea Ulbr. (Fabaceae)	146	92	5

Table 3.
Genus with the highest number of endemic taxa in Mexico with distribution in Comarca Lagunera.

compared to those of Durango. Torreón concentrates the highest number of taxa with 105 (32.7%), followed by Lerdo with 93 (29.0%) (**Table 4**). This may be due, in the first place, to the fact that the mountainous chains with greater extension and higher altitude are located mainly in these municipalities, which is conducive to the development of a large amount of endemism, as has been documented for other mountainous areas of Mexico in which the conditions of humidity and temperature allow the development of temperate vegetation [45–47], as well as in the Chihuahuan Desert [41] which shows that the diversity in this area of the country is not distributed homogeneously; and second, that the efforts to carry out floristic inventories in the semiarid zones of Coahuila have been greater; however, in the part corresponding to Durango, several inventories are in the process of being elaborated that will undoubtedly increase the knowledge of the flora.

For Comarca Lagunera, 175 taxa are recognized as scarcely distributed, that is, their representation is restricted to a single municipality, this corresponds to more than 54%, while only five taxa (1.6%) are registered in seven municipalities or more of the municipalities (**Table 5**), the above shows what in other studies has been mentioned as the arid and semiarid zones of Mexico have a high proportion of taxa with very restricted distribution [26, 48]; that in this case can be attributed possibly to the geological changes that have occurred in the Chihuahuan Desert and that have been apparently stabilized since the Miocene (15 million years) [49], so that it could be about neo-endemic taxa in the process of expansion of their populations.

In this case, 20 of the 36 taxa recognized as endemic to Comarca Lagunera are scarcely distributed (Appendix), highlighting the case of *Henricksonia mexicana* Turner, which is the only known species of the genus so far, which means that so far *Henricksonia* Turner is the only endemic genus of the flora of Comarca Lagunera. The

State/municipality	Number of taxa	Percentage with respect to the total	State/municipality	Number of taxa	Percentage with respect to the total
Durango	163	50.8	**Coahuila**	243	75.7
Lerdo	93	29.0	Torreón	105	32.7
Mapimí	57	17.8	San Pedro	87	27.1
Tlahualilo	46	14.3	Viesca	76	23.7
Rodeo	48	14.9	Francisco I. Madero	60	18.7
Nazas	18	5.60	Matamoros	18	5.6
San Pedro del Gallo	18	5.6			
General Simón Bolívar	15	4.7			
San Juan de Guadalupe	13	4.0			
Gómez Palacio	10	3.1			
San Luis del Cordero	4	1.2			

Table 4.
Number of endemic taxa of Mexico and their distribution by municipality in Comarca Lagunera.

Status	Number of taxa	Percentage
Scarcely distributed	175	54.5
Closely distributed	119	37.1
Normally distributed	22	6.9
Widely distributed	5	1.6

Table 5.
Frequency of endemic taxa of Mexico according to their distribution in the municipalities of Comarca Lagunera.

restricted distribution of the mentioned taxa means that the populations have a very small number of individuals, and therefore, it is necessary to make a greater collection effort to locate them in other places, as well as ecological studies that allow to properly characterize these populations, which will allow to have solid elements for their possible inclusion within NOM-059-ECOL-2010 [50], which is the official list in Mexico for the species that present some degree of vulnerability. Until this moment not a single taxon of the 36 taxa registered as restricted to Comarca Lagunera is included in the official list NOM-059-ECOL-2010, so it is of great importance to carry out actions that allow its conservation in the face of the increase in surface occupied for anthropogenic activities.

Coryphantha durangensis (Runge ex K. Schum.) Britton & Rose is the taxon with the widest distribution since up to now it has been registered in nine municipalities; its distribution slightly exceeds the area of Comarca Lagunera so it can be considered as quasiendemic of the region; this species is listed in NOM-059-ECOL-2010 in the status of subject to special protection (Pr), that is, "that could be threatened by factors that negatively affect its viability, which is why the need to promote its recovery and conservation is determined, or the recovery and conservation of populations of associated species," likewise this species appears in the red list of the International Union for the Conservation of Nature (IUCN) in the category of Least Concern (LC) [51]. Another 23 taxa are listed in NOM-059-ECOL-2010 and 36 taxa in the red lists of the IUCN (**Table 6**) among which *Echinocereus mapimiensis* E.F. Anderson, W. Hodgs. & P. Quirk is the only taxon restricted to the region that is included in the IUCN in the category of vulnerable (VU) [52]; *Turbinicarpus mandragora* (Fric ex A. Berger) A.D. Zimmerman and *Mammillaria pennispinosa* Krainz stand out and are distributed in Coahuila and Durango, respectively, and which according to the IUCN are in the category of critically endangered (CR) [53, 54].

The curve of accumulation of species (**Figure 6**) and the nonparametric estimator of Chao2 calculated show that the richness of endemic taxa of Mexico present in Comarca Lagunera is well known, since the result of the Chao2 estimator indicates that it is necessary to locate only eight taxa, and these data agree with the results of González-Elizondo et al. [5] for the flora of northern Mexico who mention that the flora of this region is practically well known and there are few species to be discovered; however, this value should be considered with caution since there are several sites, mainly those located in the higher mountain areas where access is complicated, in addition to the riparian zones that have not yet been documented. Likewise, the knowledge of the taxa with restricted distribution to the limits of Comarca Lagunera used in the present study has increased in the last 10 years in which 8 out of the 36 endemic taxa have been described (Appendix).

Among the vegetation types where the endemic taxa are distributed, the rosetophile desert scrub stands out, in which 192 taxa are distributed, followed by the microphile desert scrub with 147 and chaparral with 76 (**Table 7**). Furthermore, a considerable number of taxa (81) are in what has been classified as agricultural areas, so the risk of losing this part of biodiversity is considerably high; however, many of these areas have been abandoned due to the decrease in the sale prices of some products, the lack of water, the advance of desertification, and among other factors [55]. Although due to the high technology use for agriculture and the opening of sites for the breeding of various types of livestock, it cannot be ruled out that some time they are enabled again for this type of activities. A similar situation occurs in the areas that have been classified as urban in which the presence of 45 taxa has been registered, most of these localities do not present a high population density; however, the opening of new industries mainly those dedicated to mining and food production promotes the change of land use, which puts the conservation of the region's biodiversity at risk [56].

In the phenogram obtained from the cluster analysis, from the similarity matrix, three groups can be observed (**Figure 7**). In the first group appear the desert scrubs, both microphile and rosetophile, together with the chaparral and the submontane scrub, the second group is formed by the crasicaule scrub and the natural grassland, while the third group is formed by the halophile grassland, gypsophile vegetation, halophile xerophile vegetation, and vegetation of sandy deserts. It was found that the values of similarity between the different vegetation types are very low, since the microphile and rosetophile desert scrub, which are the vegetation types with the highest level of similarity, share only 41% of the taxa. The low similarity values obtained mean that the taxa represented in Comarca Lagunera have very marked preferences for the environmental conditions that prevail in each of the sites, that is, the beta diversity is very high for the set of species included in the analysis. In analysis were excluded vegetation of gallery, agricultural areas, and urban zones.

The territory of Comarca Lagunera was divided into 535 grid cells, and of these, 206 grid cells include at least one record. The average number of grid cells occupied by taxa is 3, the median is 2, and the mode is 1 (140 taxa), that is, 43.6% of the taxa have a very small distribution within Comarca Lagunera. The taxa distributed in more grid cells were *Coryphantha durangensis* (21), *Randia pringlei* (S. Watson) A. Gray (20), and *Calanticaria brevifolia* (Greenm.) E.E. Schill. & Panero (18), none of these three species restricts its distribution to Comarca Lagunera.

Taxon	**Category IUCN**	**Category NOM-059-ECOL-2010**
Agave victoriae-reginae T. Moore	LC	P
Ariocarpus kotschoubeyanus (Lem.) K. Schum.	NT	Pr
Astrophytum myriostigma Lem.	LC	A
Corynopuntia bulbispina (Engelm.) F.M. Knuth	EN	
Corynopuntia moelleri (A. Berger) F.M. Knuth	LC	
Corynopuntia vilis (Rose) F.M. Knuth	LC	
Coryphantha delaetiana (Quehl) A. Berger	LC	
Coryphantha echinus (Engelm.) Britton & Rose	LC	
Coryphantha poselgeriana (A. Dietr.) Britton & Rose	LC	A
Coryphantha gracilis L. Bremer & A. B. Lau	LC	P
Coryphantha durangensis (Runge ex K. Schum.) Britton & Rose	LC	Pr
Coryphantha longicornis Boed.	LC	
Coryphantha pseudonickelsiae Backeb.	LC	
Cylindropuntia imbricata subsp. *cardenche* (Griffiths) U. Guzmán	LC	
Cylindropuntia anteojoensis (Pinkava) E.F. Anderson	VU	Pr
Dalea melantha S. Schauer	LC	
Dalea melantha var. *pubens* Barneby	LC	
Dyssodia pinnata (Cav.) B.L. Rob.	LC	
Echinocereus mapimiensis E.F. Anderson, W. Hodgs. & P. Quirk	VU	
Echinocereus primolanatus A.F. Schwarz ex N.P. Taylor	LC	
Echinocereus stramineus subsp. *occidentalis* (N.P. Taylor) N.P. Taylor	LC	
Echinomastus unguispinus subsp. *durangensis* (Runge) U. Guzmán		A
Ephedra compacta Rose	LC	
Ferocactus pilosus (Galeotti ex Salm Dyck) Werderm.	LC	Pr
Fouquieria shrevei I.M. Johnst.		Pr
Grusonia bradtiana (J.M. Coult.) Britton & Rose	LC	
Leuchtenbergia principis Hook.	LC	A
Mammillaria coahuilensis (Boed.) Moran	EN	A
Mammillaria grusonii Runge	LC	Pr
Mammillaria lenta K. Brandegee	LC	A
Mammillaria mercadensis Patoni	LC	Pr
Mammillaria pennispinosa Krainz	CR	Pr
Mammillaria pennispinosa subsp. *nazasensis* (Glass & R. Foster) D.R. Hunt		Pr
Mammillaria stella-de-tacubaya Heese		Pr
Mammilloydia candida (Scheidw.) Buxb.	LC	
Manfreda brunnea (S. Watson) Rose		A
Manfreda potosina (B.L. Rob. & Greenm.) Rose		Pr
Opuntia leucotricha DC.	LC	
Physalis patula Mill.	LC	

Taxon	Category IUCN	Category NOM-059-ECOL-2010
Quercus vaseyana Buckley	LC	
Thelocactus bicolor subsp. *bolaensis* (Runge) Doweld		A
Thelocactus heterocromus (F.A.C. Weber) Van Oost		A
Thelocactus rinconensis subsp. *nidulans* (Quehl) Glass	LC	A
Turbinicarpus mandragora (Fric ex A. Berger) A.D. Zimmerman	CR	A

Table 6.
Endemic taxa of Mexico with distribution in Comarca Lagunera listed in some risk category. A = Threatened; CR= Critically Endangered; EN= Endangered; LC= Least Concern; NT= Near Threatened; P = In danger of extinction; Pr= Subject to special protection; VU= Vulnerable.

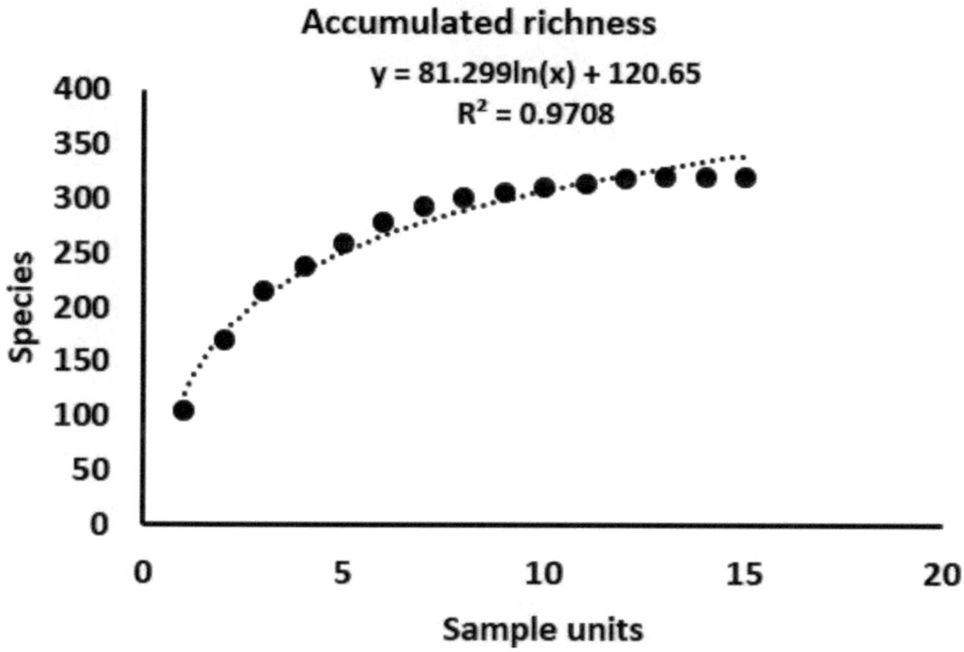

Figure 6.
Curve of accumulation of endemic taxa of Mexico with distribution in Comarca Lagunera.

The grid cells with the richest taxa were concentrated in some of the most important mountain ranges in the region. In Sierra de Jimulco (Torreón), we found two, one of them with 58 and the other one with 48 taxa, followed by Sierra de Las Delicias (San Pedro) presenting one in which 36 taxa are distributed and Sierra del Rosario (Lerdo) with one containing 31 taxa (**Figure 8**). The results obtained in the analysis of grid cells agree with the analysis at the level of municipalities mentioned above.

The grid cells with the highest WEI scores correspond with the grid cells presenting the highest taxa richness, and this is a consequence of the number of taxa rather than the values of endemism, as suggested by Crisp et al. [34]. However, the grid cells, where the highest values of the CWEI are presented, do not match with the grid cells with the greatest richness, since the highest values occur in places with few taxa. In addition, these grid cells have values of 1 for this index, in fact, there are 14 grid cells with this value, out of which 13 have a single registered taxon, and moreover, these are the only grid cells where those taxa are distributed. *S. jessicae* is the only species with restricted distribution to Comarca Lagunera and that is also

Type of vegetation	Number of taxa
Vegetation of gallery	1
Vegetation associated with bodies of water	5
Without apparent vegetation	5
Vegetation of sandy deserts	8
Induced grassland	10
Crasicaule scrub	11
Submontane scrub	19
Halophile grassland	21
Halophile xerophile vegetation	26
Gypsophile vegetation	26
Natural grassland	25
Chaparral	76
Microphile desertic scrub	147
Rosetophile desertic scrub	192

Table 7.
Distribution of taxa by type of vegetation and land use in Comarca Lagunera according to INEGI (2016).

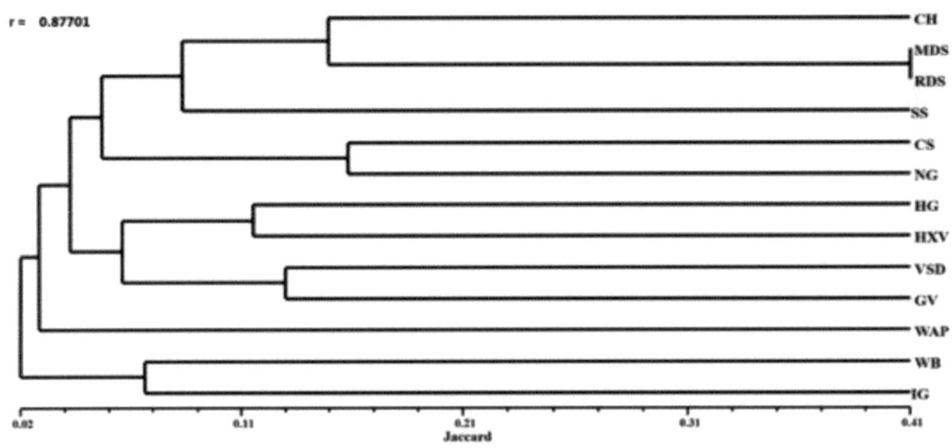

Figure 7.
Phenogram showing the relationships of the vegetation types in Comarca Lagunera based on the distribution of the endemic taxa of Mexico. CH = chaparral; CS = crasicaule scrub; GV = gypsophile vegetation; HG = halophile grassland; HXV = halophile xerophile vegetation; IG = induced grassland; MDS = microphile desert scrub; NG = natural grassland; RDS = rosetophile desert scrub; SS = submontane scrub; VSD = vegetation of sandy deserts; WAP = without apparent vegetation; and WB = vegetation associated with bodies of water.

the only taxon that has been registered in that grid cell, that is, this species can be considered as microendemic of this region of the country. If grid cells with a value of 1 are excluded from the analysis, three grid cells can be identified with a CWEI value of more than 0.750, which is the next highest value; these grid cells are located in sites that have been classified as agricultural (**Figure 9**).

The results obtained in the complementarity analysis show that seven grid cells are required to conserve at least 50% of the endemic flora of Mexico in this region (**Figure 10**); however, because a high percentage of taxa presents a distribution in

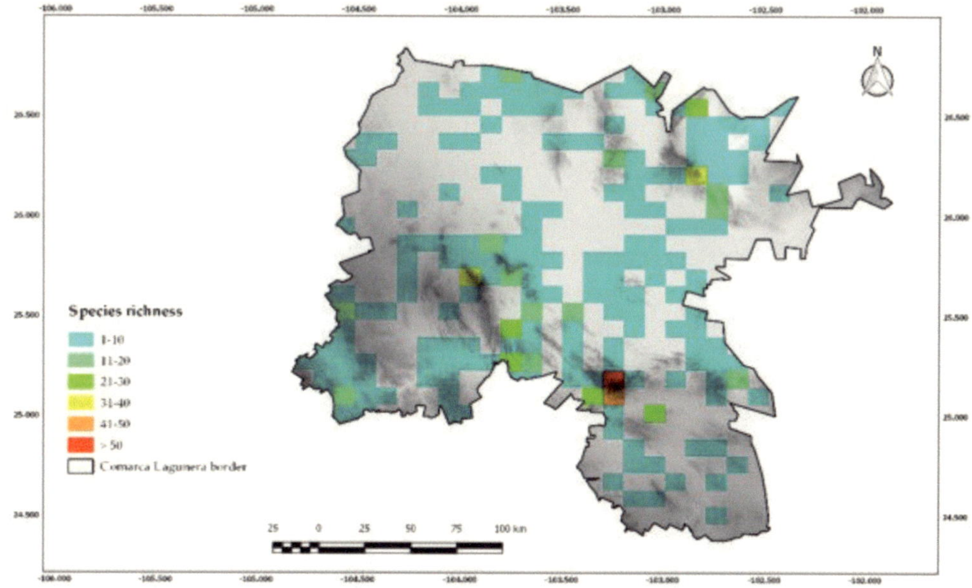

Figure 8.
Species richness in Comarca Lagunera based on the distribution in grid cells of 10 × 10 km.

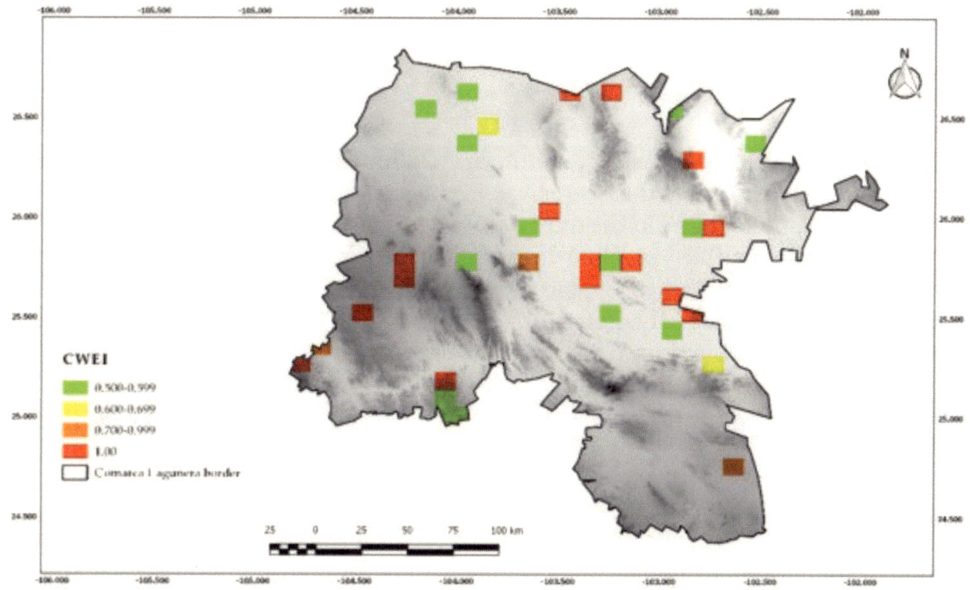

Figure 9.
Grid cells with the highest values of corrected weighted endemism index.

only one grid cell, about 90% of them is required to preserve the total of the flora included in this analysis. The previous results show that the conservation strategy that allows the protection of the endemic flora of Mexico represented in this region will be a complicated task due to the economic activities on which Comarca Lagunera depends.

Finally, the data shown here represent a first approach to the knowledge of the endemic flora present in this region of the country, and with a greater collection effort, it would probably increase. It has been reported that the causes of endemism in the

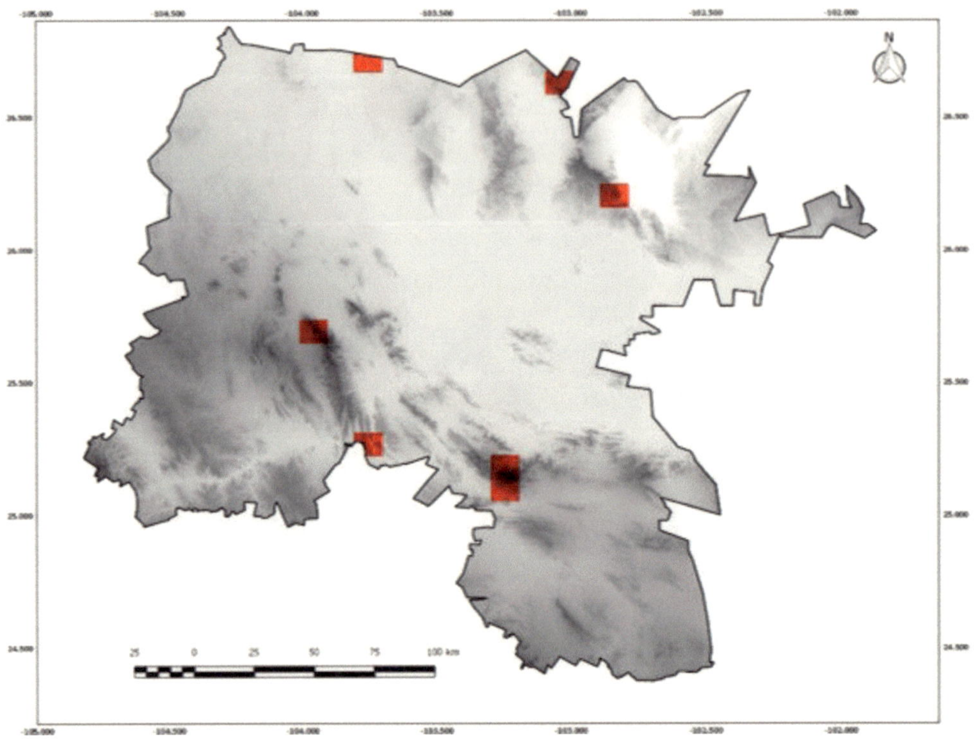

Figure 10.
Grid cells needed to conserve at least 50% of the flora in Comarca Lagunera.

arid and semiarid zones of Mexico, mainly in the Chihuahuan Desert, are frequently determined by habitat, as is the case of halophyte and gypsophyte taxa [41] and by geographic isolation mainly in the mountain areas; however, in this case, it seems that this rule is not fulfilled possibly due to the alteration of the habitats that has been caused by the different anthropogenic activities. What is a fact is that the endemic flora registered in Comarca Lagunera differs from the floristic elements found in other sites such as Cuatrociénegas and Parras de la Fuente, two localities near Comarca Lagunera, that present higher humidity and lower temperatures, since many of the known taxa of these last two sites are not represented in Comarca Lagunera [44].

4. Conclusions

The SNIB has allowed access to many researches of Mexican biodiversity in entities where there are no collections or infrastructure to house them, and they can have access to information available in scientific collections that sometimes is not available for various reasons, mainly because it is found in foreign collections, as information that is more and more strictly curated by specialists from several of the main research centers in the country; this represents one of the clearest advances in biodiversity knowledge and the cooperation of several Mexican scientific institutions, a process that began almost 30 years ago with the creation of CONABIO and later with the development of the World Net of Biodiversity Information (REMIB) that set a precedent not only in Mexico but also in various countries. Nowadays, with the development of applications in smart mobile devices such as the Naturalista platform (https://www.naturalista.mx), another huge step is taken that will allow us to know even more about the diversity of Mexico.

The present study represents an example of the value of the records of collections available in electronic media as a basis for biodiversity informatics; however, most of the available databases only present the distribution data, but lack the images of copies, so it is still necessary to review the physical specimens deposited in the collections, which have a special value since a thorough review of them and the support of the original descriptions allow the updating of the identities and the appropriate georeferencing of the specimens of a given locality [57].

In Mexico, as in other countries, conservation decisions are made taking into account political borders instead of natural criteria [32]. In this regard, the case of Comarca Lagunera is not an exception. However, in this case, efforts should be made at more local levels due to the area used in agroindustrial activities and mining exploitation. For example, some ejidos are making efforts for conservation from the scheme called payment for environmental services to which several Mexican official institutions, at different levels, contribute funds for the knowledge and subsequent conservation of biodiversity. A specific case in Comarca Lagunera is the ejido Barreal de Guadalupe, Torreón, where this type of action has been carried out successfully for a couple of years, which has impacted on the welfare of the population; however, efforts must be increased to conserve a greater surface of the semidesert zones of the north of Mexico.

Acknowledgements

We acknowledge Socorro González Elizondo and Martha González Elizondo who facilitated the consultation of specimens in the CIIDIR Herbarium. J. Jesús González Gallegos provided training for the understanding of nomenclature and descriptions in Latin.

Conflict of interest

The authors declare that they have no conflicts of interest.

Notes/thanks/other declarations

We thank Efraín Ríos Sánchez for critical read of this manuscript, for the search, and obtaining of literature with the descriptions of the species and Eliab Misael González Olvera who did an invaluable job in correcting the style of the English language.

A. List of endemic taxa of Comarca Lagunera

Taxon	**Year of description**	**Distribution**
Monocots		
Order Asparagales		
Family Asparagaceae		
Agave victoriae-reginae subsp. *swobodae* Halda	2000	General Simón Bolívar, Lerdo, Nazas, San Pedro, Torreón, Viesca

Taxon	Year of description	Distribution
Order Poales		
Family Bromeliaceae		
Hechtia mapimiana López-Ferrari & Espejo	2013	Lerdo
Eudicots		
Order Fabales		
Family Fabaceae		
Coursetia insomniifolia Lavin.	1986	San Pedro
Dalea melantha var. *pubens* Barneby.	1977	Lerdo, Torreón
Pomaria fruticosa (S. Watson) B.B. Simpson.	1998	Francisco I. Madero, Lerdo, Matamoros, San Pedro, San Pedro del Gallo, Torreón, Viesca
Order Malpighiales		
Family Euphorbiaceae		
Euphorbia cressoides M.C. Johnst.	1975	Torreón
Order Sapindales		
Family Anacardiaceae		
Cotinus chiangii (Young) Rzed. & Calderón.	1999	Lerdo
Order Malvales		
Family Malvaceae		
Batesimalva lobata Villarreal & Fryxell.	1990	Torreón
Order Brassicales		
Family Brassicaceae		
Dryopetalon stenocarpum Al-Shehbaz	2013	Mapimí, Viesca
Nerisyrenia johnstonii J.D. Bacon.	1978	Francisco I. Madero, San Pedro
Order Caryophyllales		
Family Amaranthaceae		
Atriplex monilifera S. Watson.	1874	Matamoros, Tlahualilo, Viesca
Family Polygonaceae		
Eriogonum henricksonii Reveal	1989	San Pedro
Family Caryophyllaceae		
Drymaria elata I.M. Johnst.	1940	Francisco I. Madero, San Pedro
Drymaria jenniferae Villarreal & A.E. Estrada	2008	Viesca
Family Cactaceae		
Corynopuntia halophila D. Donati	2017	Francisco I. Madero, San Pedro
Echinocereus mapimiensis E.F. Anderson, W. Hodgs. & P. Quirk	1998	Tlahualilo
Order Cornales		
Family Loasaceae		

Taxon	Year of description	Distribution
Eucnide durangensis H.J. Thomps. & Powell	1981	Lerdo, Rodeo, Torreón
Order Ericales		
Family Primulaceae		
Samolus dichondrifolius Channell	1958	San Pedro
Order Gentianales		
Family Rubiaceae		
Hedyotis teretifolia (Terrell) G.L. Nesom	1988	Francisco I. Madero, San Pedro
Machaonia pringlei A. Gray	1886	Torreón, Viesca
Family Apocynaceae		
Matelea greggii (Vail) Woodson	1941	Mapimí
Order Boraginales		
Family Boraginaceae		
Lithospermum jimulcense I.M. Johnst.	1952	Torreón, Viesca
Order Solanales		
Family Solanaceae		
Solanum johnstonii M. D. Whalen	1976	Lerdo, Torreón, Viesca
Order Lamiales		
Family Plantaginaceae		
Mabrya coccinea (I.M. Johnst.) Elisens	1985	Francisco I. Madero, Matamoros, San Pedro
Family Scrophulariaceae		
Leucophyllum coahuilensis Henr.	2004	Francisco I. Madero
Family Lamiaceae		
Salvia jessicae B.L. Turner	2013	Francisco I. Madero
Family Acanthaceae		
Justicia decurvata Hilsenb.	1990	Lerdo
Ruellia jimulcensis Villarreal.	1998	Lerdo, Matamoros, Torreón
Siphonoglossa durangensis Henr. & Hilsenb.	1979	Lerdo
Siphonoglossa linearifolia Henr. & Hilsenb.	1979	Lerdo
Order Asterales		
Family Asteraceae		
Haploesthes hintoniana B.L. Turner	2013	Francisco I. Madero
Henricksonia mexicana B.L. Turner	1977	Lerdo, Nazas, Torreón
Flaveria intermedia J.R. Johnst.	1903	Tlahualilo
Gaillardia candelaria B.L. Turner var. *candelaria*	2007	San Pedro
Gaillardia candelaria var. *mikemoorei* B.L. Turner	2013	Francisco I. Madero
Marshalljohnstonia gypsophila Henrickson.	1977	San Pedro

Author details

Alberto González-Zamora[1*] and Rebeca Pérez-Morales[2]

1 Laboratory of Evolutionary Biology, Biological Science Faculty, Juárez University of the State of Durango, Gómez Palacio, Durango, Mexico

2 Laboratory of Cellular and Molecular Biology, Faculty of Chemical Sciences, Juárez University of the State of Durango, Gómez Palacio, Durango, Mexico

*Address all correspondence to: agzfc@ujed.mx

References

[1] La Salle J, Williams KJ, Moritz C. Biodiversity analysis in the digital era. Philosophical Transactions of the Royal Society B. 2016;**371**:20150337. DOI: 10.1098/rstb.2015.0337

[2] Peterson AT, Ingenloff K. Biodiversity informatics training curriculum, versión 1.2. Biodiversity Informatics. 2016;**10**:65-74. DOI: 10.17161/bi.v11i0.5008

[3] Jiménez R, Koleff P. La informática de la biodiversidad: una herramienta para la toma de decisiones. In: Capital natural de México, vol IV: capacidades humanas e institucionales. México: CONABIO; 2016. pp. 143-195

[4] Daly M, Herendeen PS, Guralnick RP, Westneat MW, McDade L. Systematics Agenda 2020: The misión evolves. Systematic Biology. 2012;**61**:549-552. DOI: 10.1093/sysbio/sys044

[5] González-Elizondo MS, González-Elizondo M, López-Enríquez IL, Tena-Flores JA, González-Gallegos JG, Ruacho-González L, et al. Diagnóstico del conocimiento taxonómico y florístico de las plantas vasculares del norte de México. Botanical Sciences. 2017;**95**:760-779. DOI: 10.17129/botsci.1865

[6] Henrickson J, Johnston MC. A flora of the Chihuahuan Desert Region. Los Angeles, USA: Published by J. Henrickson; 2007. p. 1695

[7] Sánchez Á. Delimitación geográfica. In: López A, Sánchez A, editors. Comarca Lagunera: procesos regionales en el contexto global. Primera ed. México: UNAM-Instituto de Geografía; 2010. pp. 15-36

[8] García E. Modificaciones al sistema de clasificación climática de Köppen. 5th ed. México: UNAM-Instituto de Geografía; 2004. p. 90

[9] Gabriel J, Pérez JL. Paisajes geográficos naturales. In: López A, Sánchez A, editors. Comarca Lagunera: procesos regionales en el contexto global. Primera ed. México: UNAM-Instituto de Geografía; 2010. pp. 75-97

[10] Challenger A. Utilización y conservación de los ecosistemas terrestres de México: Pasado, presente y futuro. Vol. 847. UNAM-Instituto de Biología: México; 1998

[11] Miranda F, Hernández E. Los tipos de vegetación de México y su clasificación. Boletín de la Sociedad Botánica de México. 1963;**28**:29-179

[12] Rzedowski J. Vegetación de México. 1st digital ed. México, Comisión Nacional para el Conocimiento y Uso de la Biodiversidad; 2006. p. 504

[13] Instituto Nacional de Estadística y Geografía (INEGI). Conjunto de datos vectoriales de Uso de Suelo y Vegetación. Escala 1:250 000. Serie VI (Capa Unión). Primera ed. México: INEGI. 2016. Available from: http://www.conabio.gob.mx/informacion/gis/ [Accessed: 14-05-2018]

[14] González F. Las comunidades vegetales de México. In: Propuesta para la unificación de la clasificación y nomenclatura de la vegetación de México. 1st ed. México: Instituto Nacional de Ecología (INE-SEMARNAT); 2003. p. 77

[15] Rivas-Martínez S, Sánchez-Mata S, Costa M. North American boreal and western temperate forest vegetation (syntaxonomical synopsis of the potential natural plant communities of North America, II). Itinera Geobotánica. 1999;**12**:5-316

[16] González-Elizondo M, González-Elizondo MS, Arrieta Y. Listados Florísticos de México. In: IX Flora de

Durango. Primera ed. México: UNAM-Instituto de Biología; 1991. p. 167

[17] Villarreal-Quintanilla JA. Listados Florísticos de México. In: XXIII Flora de Coahuila. Primera ed. México: UNAM-Instituto de Biología; 2001. p. 138

[18] García-Arévalo A. Vascular plants of the Mapimí biosphere reserve, Mexico: A checklist. Sida. 2002;**20**:798-807

[19] Villarreal-Quintanilla JA, Encina-Domínguez JA. Plantas vasculares endémicas de Coahuila y algunas áreas adyacentes, México. Acta Botanica Mexicana. 2005;**70**:1-46

[20] González-Elizondo MR, Galván-Villanueva IL, López-Enriquez L, Reséndiz-Rojas L, MS G-E. Agaves-magueyes, lechuguillas y noas- del Estado de Durango y sus alrededores. Primera ed. México: CIIDIR Instituto Politécnico Nacional-CONABIO; 2009. p. 163

[21] Hernández HM, Gómez-Hinostrosa C. Mapping the cacti of Mexico. Succulent Plant Research. 2011;**7**:1-128

[22] Sánchez J, Estrada-Castillón E, Arias S, Muro G, García-Aranda M, García-Morales LJ. Diversidad cactoflorística de la zona árida y semiárida de Durango, México. Interciencia. 2014;**39**:794-802

[23] Hernández HM, Gómez-Hinostrosa C. Mapping the cacti of Mexico. Part II Mammillaria. Succulent Plant Research. 2015;**9**:1-189

[24] Montelongo-Landeros M, Alba-Ávila JA, Romero-Méndez U. García-De la Peña C. Pteridophytas de las sierras El Sarnoso y Mapimí en Durango, México. Revista Mexicana de Biodiversidad. 2015;**86**:448-456. DOI: 10.1016/j.rmb.2015.04.029

[25] Valdés J. Gramíneas de Coahuila. Primera ed. México: CONABIO; 2015. p. 556

[26] González-Elizondo M, González-Elizondo MS, González-Gallegos JG, Tena-Flores JA, López-Enriquez IL, Ruacho-González L, et al. Updated checklist and conservation status of Cactaceae in the state of Durango, Mexico. Phytotaxa. 2017;**327**:103-129. DOI: 10.11646/phytotaxa.327.2.1

[27] Villaseñor JL. Checklist of the native vascular plants of Mexico. Revista Mexicana de Biodiversidad. 2016;**87**: 559-902. DOI: 10.1016/j.rmb.2016.06.017

[28] The Angiosperm Phylogeny Group. An update of the Angiosperm Phylogeny Group classification for the orders and families of flowering plants: APG III. Botanical Journal of the Linnean Society. 2009;**161**:105-121. DOI: 10.1111/j.1095-8339.2009.00996.x

[29] SEMARNAT-CONANP. *'182ANP_Geo_ITRF08_Noviembre_2017'*, ed: 2017. Secretaría de Medio Ambiente y Recursos Naturales, Comisión Nacional de Áreas Naturales Protegidas. 2017 Ciudad de México, México: 2017. Available from: http://www.conabio.gob.mx/informacion/gis/ [Accessed: 14-05-2018]

[30] CONABIO. *'Áreas Naturales Protegidas Estatales, Municipales, Ejidales y Privadas de México* 2015, ed: 1. Comisión Nacional para el Conocimiento y Uso de la Biodiversidad. Distrito Federal, Tlalpan. 2015. Available from: http://www.conabio.gob.mx/informacion/gis/ [Accessed: 14-05-2018]

[31] Team QGIS Development. QGIS Geographic Information System. Open Source Geospatial Foundation Project 2018. Available from: https://www.qgis.org/

[32] Dávila-Aranda P, Lira R, Valdés-Reyna V. Endemic species of grasses in Mexico: A phytographic approach. Biodiversity and

Conservation. 2004;**13**:1101-1121. DOI: 10.1023/B:BIOC.0000018147.54695.b3

[33] Rohlf FJ. NTSYSpc. Numerical Taxonomy and Multivariate Analysis System. Versión 2.0. Exeter Software, Nueva York; 1998

[34] Crisp MD, Laffan S, Linder HP, Monro A. Endemism in the Australian flora. Journal of Biogeography. 2001;**28**:183-198. DOI: 10.1046/j.1365-2699.2001.00524.x

[35] Linder HP. Plant diversity and endemism in sub-Saharan tropical Africa. Journal of Biogeography. 2001;**28**:169-182. DOI: 10.1046/j.1365-2699.2001.00527.x

[36] Contreras-Medina R, Luna-Vega I. Species richness, endemism and conservation of Mexican gymnosperms. Biodiversity and Conservation. 2007;**16**:1803-1821. DOI: 10.1007/s10531-006-9072-3

[37] Santa Anna Del Conde H, Contreras-Medina R, Luna-Vega I. Biogeographic analysis of endemic cacti of the Sierra Madre oriental, Mexico. Biological Journal of the Linnean Society. 2009;**97**:373-389. DOI: 10.1111/j.1095-8312.2009.01212.x

[38] Sosa V, de-Nova A. Endemic angiosperm lineages in Mexico: Hotspots for conservation. Acta Botanica Mexicana. 2012;**100**:293-315

[39] Contreras-MacBeath T, Brito M, Sorani V, Goldspink C, McGregor G. Richness and endemism of the freshwater fishes of Mexico. Journal of Threatened Taxa. 2014;**6**:5421-5433. DOI: 10.11609/JoTT.o3633.5421-33

[40] Vane-Wright RI, Humphries CJ, Williams PH. What to protect?: Systematics and the agony of choice. Biological Conservation. 1991;**55**:235-254. DOI: 10.1016/0006-3207(91)90030-D

[41] Villarreal-Quintanilla JA, Bartolomé-Hernández JA, Estrada-Castillón E, Ramírez-Rodríguez H, Martínez-Amador SJ. El elemento endémico de la flora vascular del Desierto Chihuahuense. Acta Botanica Mexicana. 2017;**118**:65-96. DOI: 10.21829/abm118.2017.1201

[42] Donati D. *Corynopuntia halophila* (Cactaceae), a new species from Coahuila, Mexico. Plant Biosystems. 2018;**152**:386-397. DOI: 10.1080/11263504.2017.1297332

[43] Turner BL. Taxonomic overview of the Mexican species of *Salvia* sect. *Flocculosae* (Lamiaceae). Phyton. 2013;**36**:1-11

[44] JA V-Q, AE E-C. A new species of Drymaria (Caryophyllaceae) from northeastern Mexico. Brittonia. 2008;**60**:329-331. DOI: 10.1007/s12228-008-9028-x

[45] Ruacho-González L, González-Elizondo MS, González-Elizondo M, López-González C. Diversidad florística en cimas de la Sierra Madre Occidental. Botanical Science. 2013;**91**:193-205

[46] Rzedowski J. Catálogo preliminar de plantas vasculares de distribución restringida a la Sierra Madre oriental. In: Flora del Bajío y de Regiones Adyacentes. Fascículo complementario XXXI. México: Instituto de Ecología del Bajío; 2015. p. 36

[47] Suárez-Mota E, Villaseñor JL, Ramírez-Aguirre MB. Sitios prioritarios para la conservación de la riqueza florística y el endemismo de la Sierra Norte de Oaxaca, México. Acta Botanica Mexicana. 2018;**124**:49-74. DOI: 10.21829/abm124.2018.1296

[48] Hernández HM, Gómez-Hinostrosa C, Hoffmann G. Is geographical rarity frequent among the cacti of the Chihuahuan Desert? Revista Mexicana de Biodiversidad. 2010;**81**:163-175

[49] Wilson JS, Pitts JP. Illuminating the lack of consensus among descriptions of earth history data in the North American deserts: A resource for biologists. Progress in Physical Geography. 2010;**34**:419-441. DOI: 10.1177/0309133310363991

[50] SEMARNAT. NORMA Oficial Mexicana NOM-059-SEMARNAT-2010, Protección ambiental-Especies nativas de México de flora y fauna silvestres-Categorías de riesgo y especificaciones para su inclusión, exclusión o cambio-Lista de especies en riesgo. Diario Oficial de la Federación. 15 de diciembre de 2010. Available from: http://dof.gob.mx/nota_detalle.php?codigo=5173091&fecha=30/12/2010

[51] Dicht RF, Lüthy AD, Goettsch BK. *Coryphantha durangensis* (amended version of 2013 assessment). The IUCN Red List of Threatened Species; 2017. e.T62361A121439780. DOI: 10.2305/IUCN.UK.2017-3.RLTS.T62361A121439780

[52] Fitz Maurice B, Sotomayor M, Fitz Maurice WA., Hernández HM, Smith M. *Echinocereus mapimiensis* (amended version of 2013 assessment). The IUCN Red List of Threatened Species; 2017. e.T151774A121442820. DOI: 10.2305/IUCN.UK.2017-3.RLTS.T151774A121442820.en

[53] Fitz Maurice B, Gómez-Hinostrosa C, Fitz Maurice WA, Hernández HM. *Turbinicarpus mandragora*. The IUCN Red List of Threatened Species; 2013. e.T40982A2949250. DOI: 10.2305/IUCN.UK.2013-1.RLTS.T40982A2949250

[54] Fitz Maurice WA, Fitz Maurice B. Mammillaria pennispinosa. The IUCN Red List of Threatened Species; 2013. e.T40846A2936587. DOI: 10.2305/IUCN.UK.2013-1.RLTS.T40846A2936587

[55] Miranda R. Caracterización de la producción del cultivo de algodonero (*Gossypium hirsutum* L.) en la Comarca Lagunera. Revista Mexicana de Agronegocios. 2008;**23**:696-705

[56] Muro-Pérez G, Sánchez-Salas J, Alba-Ávila JA. Desarrollo agroindustrial: Reseña y perspectiva en la Comarca Lagunera, México. Revista Chapingo Serie Zonas Áridas. 2012;**11**(1):1-7

[57] Veiga AK, Cartolano EA Jr, Saraiva AM. Data quality control in biodiversity informatics: The case of species ocurrence data. IEEE Latin America Transactions. 2014;**12**:683-693. DOI: 10.1109/TLA.2014.6868870

Chapter 5

Mexican Indigenous Species with Agroecological Uses

Arellano-Perusquía Abraham, Bañuelos-Hernández Bernardo, Urquieta-Ramírez Luz, Albarrán-Tamayo Froylán, Aguilar-Ruiz Carlos Agustín, González-Márquez Marco Antonio, Junquera-Martínez Sergio and Cruz-Hernández Andrés

Abstract

Mexico is considered one of the twelve megadiverse countries, and they together harbor 60–70% of the biodiversity from the planet. Mexico with Brazil, Colombia, and Indonesia occupied the principal positions. In Mexico, there are almost 50,000 plant species recognized now. Several of these plants are used since the pre-Hispanic age in religious ceremonies and medicinal treatments, but their potential in agroecology has not been exploited. There is not much information related to these alternative crops and their possible uses in agriculture. In this chapter, we will describe the main characteristics of the plant and its possible uses in agroecology.

Keywords: ancient plants, biocides, conservation, *Heliopsis*, *Tagetes*

1. Introduction

Biodiversity determines the interaction between organisms, their environment, and the evolution through the time [1]. It has been estimated that the number of species in the world is between 5 and 30 million [2], although it is difficult to establish these numbers with precision.

Because its large territory and microclimates, Mexico harbors a high number of plant species and genera, becoming one of the countries with the highest diversity in plants, reptiles, and mammals [3]. Mexico is one of the megadiverse countries and is estimated to harbor 10% of the planet flora [3]. It was calculated that Mexico has approximately 24,000 plant species, 535 mammal species, 804 reptile species, 361 amphibians, and 1107 birds, in accordance with Mexico's information at the Biological Diversity Agreement.

Several species are Mexicans exclusively; 50 to 60% of the known plant species are in the Mexican territory: 21 pine species, 146 agaves, and 175 cacti. In this chapter, we present some indigenous Mexican plant species and their agroecological uses that could represent interest for people.

2. Methodology

Several sources were consulted in order to obtain the information for this document that included databases, bibliographical sources, interviews with specialists and countrymen, and available information in the official catalogs of the plants.

3. Results

3.1 *Tagetes* genera

3.1.1 Botanical description

Asteraceae family is distributed mainly at the American continent, and 10% of the plant species is located in Mexico [4, 5]. The genus *Tagetes* is represented by 58 endemic species from America, and in Mexico there are 36 species. Three of them are introduced from Central and South America: *T. microglossa*, *T. minuta*, and *T. terniflora*. **Table 1** shows the taxonomical description for the genus *Tagetes*.

The *Tagetes* plants have pinnate leaves and stems with a growth from 1 cm to 2 m high, depending on the species. Their flowers are ligulated with sizes from 1 to 2 cm and growth around the plant in a radial form with a diameter of 4 to 6 cm. Flowers show red, orange, yellow, and white colors [6]. **Table 2** shows some of the characteristics of the plants from the genus *Tagetes* [6, 7].

3.1.2 Uses of the genus Tagetes

The *Tagetes* secondary metabolites give the plant the medicinal properties and the biological effects against several microorganisms (bacteria, protozoa, and fungi), insects, and nematodes that provoke some damage to plants; some of them are dangerous to animals and humans. Some species, such as *T. erecta*, *T. patula*, and *T. lucida*, produce pigments used in the agroindustry for the coloring of food products and as nutraceuticals for its antioxidant properties. **Table 3** included some of the species of the genus *Tagetes* cultivated in Mexico.

Kingdom	*Plantae*
Subkingdom	*Viridiplantae*
Infrakingdom	*Streptophyta*
Division	*Tracheophyta*
Subdivision	*Spermatophytina*
Infradivision	*Angiospermae*
Class	*Magnoliopsida*
Superorder	*Asteranae*
Order	*Asterales*
Family	*Asteraceae*
Tribe	*Tageteae*
Genus	*Tagetes*

Table 1.
Taxonomical classification for the genus Tagetes from the Integrated System of Information (SIIT, 2012).

Specie	Stem high (cm)	Petal num ber	Flower color	Flower sizes (mm)	Head flower color	Germination
T. lunulata	30-80	20-35	yellow	7-10	green	annual/herbal
T.foetidissima	20-80	4-8	yellow green	5-6.5	green	annual/herbal
T. multiflora	20	7-14	yellow	2-4	purple	annual/herbal
T.coronopifolia	50	5-20	yellow	2-4	purple	annual/herbal
T. linifolia	20-40	40-60	yellow, orange, red	5-12	green	perennial/herbal
T. lemmonii	30-100	30-60	yellow	10-15	green	perennial/herbal
T. stenophylia	50-100	3-60	yellow	8-12	radial	perennial/herbal
T. patula	20-85	56-70	yellow, orange, red	8-13	green	annual/herbal
T. persicaefolia	100	7-8	yellow	8-12	green	annual/herbal
T. erecta	40-60	40-90	yellow, orange	10-15	green	annual/herbal
T. moorei	8-40	10-20	yellow	8-10	green	annual/herbal
T. parryi	30-50	50-60	yellow	12-20	green	annual/herbal
T. minuta	150	3-5	yellow	0.5-1	green	annual/herbal
T. terniflora	160	3-5	yellow	0.5-1	red	annual/herbal
T. lúcida	100	5-8	yellow	6-7	green	perennial/herbal
T. arenícola	30-60	5-6	Yellow, orange	6-7	green	annual/herbal
T. linifolia	10-40	3-9	yellow	3-3.5	green	annual/herbal
T. micrantha	10-50	4-10	yellow	3-3.5	green	annual/herbal
T. subulata	10-50	10-20	Yellow, yellow	7-10	brown	annual/herbal

Table 2.
Morphological characteristics of some species of Tagetes *[6, 7].*

T. arenicola	*T. jaliscensis*	*T. moorei*	*T. pringlei*
T. coronopifolia	*T. lacera*	*T. mulleri*	*T. remotiflora*
T. epapposa	*T. lemmonii*	*T. nelsonii*	*T. stenophylla*
T. elongata	*T. linifolia*	*T. oaxacana*	*T. subulata*
T. erecta	*T. lunulata*	*T. palmerii*	*T. subvillosa*
T. filifolia	*T. lucida*	*T. parryi*	*T. tenuifolia*
T. foetidissima	*T. micrantha*	*T. patula*	*T. terniflora*
T. hartwegii	*T. minuta*	*T. persicaefolius*	*T. trirradiata*

Table 3.
Species del genus Tagetes *cultivated in Mexico [6].*

There are several reports related to the biocide effect of some species of *Tagetes* over plant parasites that affect the yield in potato and tomato [8–13]. The reported species are *T. patula*, *T.* minuta, and *T. erecta* that reduce the presence of parasites such as *Rh. Sanguineus*, *Phlebotomus duboscqi*, *Pratylenchus penetrans*, and *Meloidogyne incognita* [14].

The biological and chemical knowledge of *Tagetes* is not exhaustive, although in some species, such as *T. erecta* and *T. minuta*, the research had been extensive worldwide, because the commercial exploiting of the pigments and the essential oils results in the development of a very promising agroindustry in Africa, China, India, the USA, and Peru.

The main characteristic of *Tagetes* is that their species are aromatic that is related to the medicinal approach since the Mexican pre-Hispanic age. This knowledge had been conserved to the actual times in the traditional medicine.

3.1.3 Historical context and importance of the genus Tagetes

The plants of the genus *Tagetes* had been used since the pre-Hispanic age by the nuatlaca people. Its uses were described at the *Florentino Codex* by Fray Bernardino de Sahagún in the sixteenth century. An example is *T. lucida*, known in Nahuatl as yauhtli. It had been used as air freshener and ornamental in fire and rain celebrations and also was used as medicine for the treatment of fever and ulcers. Another example is *T. tenuilora*, used as an herbal food condiment.

The indigenous people also used *T. erecta* known as flor de Cempoalxóchitl used at the temples and houses in the sun and dead people (Dia de Muertos) festivities [6]. Now, flor de Cempoalxóchitl is still used in the traditional medicine as antiparasite and in the treatment of stomach, bazo, and liver diseases. Because of the high content of carotenoids, especially lutein, the flowers of *T. erecta* are used at the food industry as additive in the animal food for aves de corral, fishes, and crustaceans for the pigmentation of their skin, making them very attractive for their consumption [14].

The carotenoids of Cempoalxóchitl and other species of *Tagetes* are used in the treatment of human diseases. The xanthophylls and β-carotene (vitamin A precursor) have an antioxidant activity to protect from the damage provoked by the free radicals, responsible of several degenerative diseases, such as arteriosclerosis, arthritis, and carcinogenesis.

Because the use of pesticides in the crops culture, there are some problems in those and their products: high cost, plague resistance, and pollution. Now some strategies to avoid these problems are being looked for. *Tagetes* such as *T. minuta*, *T. patula*, and *T. erecta* produce terpenes and thiophenes with pesticidal and acaricidal capacity that could be used as extracts or in co-cultivation for the control of plagues that produce some damage in several cultures [8–13].

3.1.4 Tagetes production in Mexico

In Mexico there are 859 ha cultivated with *T. erecta*, 69.2% are grown in Puebla state, 10% in State of Mexico, 2.3% Michoacán state, and 0.33% in Mexico City. The rest of the production is in the states of Oaxaca, Guerrero, Hidalgo, Morelos, San Luis Potosí, Tlaxcala, and Durango. The year production reaches 6598 tons of flower with a value of 16 million Mexican pesos. The production is mainly for ornamental in the "Dia de Muertos" celebration [7]. *T. patula* is also used as ornamental; however, there is not available data for its production. The rest of the species are growing in a wild way [6].

The production system is through transplant. The plantlets are developed in seedlings, and they are transplanted to the field when they are 10 cm high (20 to 45 days). In the field, the plantlets continue their development until flowering until the end of the cycle (110 days). The flower yields are variable in the culture; they are from 12 to 30 tons of flower/ha, in a maximal of seven cuts [6].

3.1.5 Taxonomical classification and geographical distribution of the Tagetes

The genus *Tagetes* is distributed from the southwest of the USA to the south of Argentina. In Mexico, 32 from the 56 discovered species had been found, and they are localized in several states around the country. Several authors classified the *Tagetes* species based on its morphological characteristics; however, these species show very similar characteristics, causing confusion and erroneous taxonomical classifications in several species. There are different names for the same species, or also some plants considered with independent species are now recognized as members of the same species, i.e., *T. tenuifolia* and *T. peduncularis*, which have been renamed as *T. lunulata* [6]. There is also the case of *Addenopapus persicaefolius*, incorporated to the genus *Tagetes* by molecular evidence [6].

Since 2003, in Mexico, a project was initiated for the compilation and analysis of the available information of the species of the genus *Tagetes*; its center of origin and diversity is Mexico. **Table 3** shows a list of 32 species found in Mexico [6].

3.2 Cactaceae family

3.2.1 Geographical distribution

The most notable plants in the Mexican arid landscapes, including the *Agave*, mesquites, and *Yucca*, are the Cactaceae family [15, 16]. The Cactaceae family is native from America. It is found from the south of Canada to Argentina; Mexico is harboring the highest diversity of species and a high index of endemism (78%) [17].

The distribution in Mexico includes several environments, such as Zapotitlán de las Salinas in Tehuacan, at Puebla; the Tomellin Canyon, the Sierra Mixteca in Oaxaca; the ravines in Metztitlan, Hidalgo; the Rio Balsas Basin; the Infiernillo ravines in Querétaro; the Potosino-Zacatecano plateau; and the Baja California Peninsula [15].

The Cactaceae plants are distributed all around the country, and they are mainly concentrated at the arid and semiarid lands of San Luis Potosí, Zacatecas, Aguascalientes, Querétaro, Hidalgo, and Chihuahua [17]. The Querétaro and San Luis Potosí are the center of distribution of a high number of species, because their geographical and climate isolation from the rest of the country have originated several members of this plant family. The phytogeographical and biological importance of this area is due to the species that grow, not present in any other regions of the planet [18].

3.2.2 Botanical description

The Cactaceae are perennial plants, succulents, terrestrial, geophytes, thorny, arborescent, or epiphytes. Their epidermis is glabra or pubescent; they develop fibers and tuberculous roots. The stems are globose, cylindrical, columnar, or flattened, with tubers or moms (mamilas) in a spiral disposition, with a constant number in each species; there are some other structures such as wings, ribs, or flatted and green phylloclades. The leaves are laminar in the primitive genera and absent in other genera. The areoles harboring reproductive and vegetative meristems, such as axillar buds with spines and glands, can produce flowers or leaves [18].

The flowers are zoophilous, mainly sessile, isolated, and rarely grouped as inflorescence, such as panicles or cimosas (*Pereskia*) or terminal (*Pterocactus*); they are hermaphrodites, sometimes unisexual, and almost actinomorphic with betalains. The pericarpel, tissue from the stem located around the ovary or the receptacle, is glabro or covered with areoles in a spiral way. They are normally covered by scales, and they used to be integrated with the perianth, showing glochids and spines. The perianth is formed by external or internal segments of petaloids, associated in series in spiral form. The superior ovary is found in a unique species *Pereskia aculeata,* although in general the ovary is inferior or semi-inferior, with 3 to 20 unilocular carpels. The flowers have a unique style, and the stigmas develop several lobules with different numbers depending on the species. The ovules are characterized by their long funicles, concrescent in fascicles [6].

The fruit could be flesh, dried or semidried, indehiscent or dehiscent, and sometimes glabro or with areoles harboring spines. The succulence is due to the funicles that accumulate sugar during fruit ripening.

The seeds differentiate in a perisperm as a food storage system for the embryo; the testa could be thin or thick with or without a ring (strophiole) originated at the funicle. The cotyledons are foliaceous at the primitive genera and reduced or primitive in mainly all the species [6].

3.2.3 Endangered species

In Mexico, the Cactaceae are endemic plants that play an important role in a biological, social, and economic point of view. However, the plantlets and the adult plants have been stolen from their habitat for commercial proposes in the USA, Japan, and several European countries [15, 16]. In the last years, there is an increase in the demand for the cactus plants, mainly in the species considered as rare or unique; these plants are commercialized by national or international collectors. These plants are collected in situ, increasing the pressure over the wild populations and affecting the rate of natural reproduction taken some species to the category of endangered species [15, 16]. Now 255 taxons of Cactaceae are considered endangered species as described by Norma Oficial Mexicana-059-ECOL-2001. *The Red Book* (IUCN) and the CITES appendix include 65 and 41 taxons, respectively, as endangered species [19].

3.2.4 Botanical uses of the Cactaceae family

Ortegocactus macdougallii (Alexander)

It is important as ornamental and appreciated by the collectors because of its flowers and stems; this plant helps in the soil restoration and erosion control and is a very good rain harvester.

Turbinicarpus ysabelae (Schlange) (John & Riha)

It is important as ornamental, its fruits are a good source for food to the local fauna, and this plant helps to control the soil structure and erosion and harvest polluted gases.

Turbinicarpus schmiedickeanus ssp. jauernigii (G. Frank) (D.R. Hunt)

It is important as ornamental and in the control of soil erosion; this plant is used as source of phytochemicals (alkaloids) for medicinal and ludic purposes.

Turbinicarpus swobodae (Diers & Esteves)

It is used as ornamental and helps in the control of soil erosion. This plant helps to harvest the water from the rain and in the entrapment of environmental pollutants.

Astrophytum asterias (Zucc.) (Lem)

These plants produce secondary metabolites (alkaloids) and are used in religious ceremonies and in the traditional medicine.

Mammillaria hernandezii (Glass *et* Foster)

It is used as ornamental use and helps in the control of soil erosion.

3.3 Chilcuague (*Heliopsis longipes*)

3.3.1 Origin and common uses

Heliopsis longipes, which common name is “chilcuague,” is a Mexican herbaceous plant which is found in a mountainous area where the states of.

San Luis Potosí, Guanajuato, and Querétaro meet. Pre-Hispanic Mexicans employed the extracts from the root system of this plant to heal diverse pains such as toothache, earache, and headache [20]; it was recognized for its pungent taste and for causing numbing and salivation. Three centuries later, in 1948, Elbert L. Little Jr. published a small description of *Heliopsis longipes* and suggests the potential use of the plant as an insecticidal [21].

Despite its importance the use of chilcuague roots was marginal and restricted to the communities of the localities where the plant grows; one of the main reasons for this phenomenon is the fact that the massive propagation of the plant has experienced problems, mainly in transplanting after seedling stage.

3.3.2 Agroecological potential

Alkamides are secondary metabolites comprising over 200 related compounds widely distributed in plants. They have been found in ten plant families: Aristolochiaceae, Asteraceae, Brassicaceae, Convolvulaceae, Euphorbiaceae, Menispermaceae, Piperaceae, Poaceae, Rutaceae, and Solanaceae. The Asteraceae, Piperaceae, and Rutaceae families comprise the diverse species that contain high levels of alkamides [22–24]. The general structure of alkamides originates from the condensation of an unsaturated fatty acid and an amine [25]. Although different chain length alkamides have been found in plants, most of them contain a 2*E* double bond conjugated to the amide group substituted with an *N*-isobutyl group. Recently, it has been shown that alkamides are the main compounds found in *Heliopsis longipes* roots and are responsible for diverse biological effects on bacteria, insects, plants, and humans [26, 27]. The main alkamides isolated from chilcuague roots are *N*-isobutyl-2*E*,6*Z*,8*E*-decatrienamide (afinin), *N*-isobutyl-2*E*-decenamide, and *N*-isobutyl-decanamide. The biological effects of affinine on diverse organisms have been tested:

The effect of ethanolic extract from *Heliopsis* roots and purified affinine was tested on grubs of *Aedes aegypti*, the vector for dengue transmission. Both, the extract and pure affinine, showed a high toxic effect on the insect grubs, demonstrating that the molecule responsible for this lethality is affinine. Despite the exciting role of affinine on *Aedes aegypti* mortality, further studies are most performed in order to know the specific target of affinine.

Affinine and affinine-derived molecules display a negative activity against mycelial growth of diverse phytopathogenic fungi, such as *Phytophthora infestans*, *Fusarium* spp., *Sclerotium rolfsii*, and *Verticillium* spp. Also negative effect on cell growth and division of *Escherichia coli* and *Bacillus subtilis* was shown.

Ramírez-Chávez and co-workers also investigated the effects of affinine and its derivates, *N*-isobutyl-2*E*-decenamide and *N*-isobutyl-decanamide, on plant growth and early root development of *Arabidopsis thaliana*. They found that treatments with affinine, in the range of $7x\ 10^{-6}$–$2.8\ x\ 10^{-5}$ M, enhanced the primary root growth and root hair elongation, whereas higher concentrations inhibited primary root growth that is related to a reduction in cell proliferating activity and cell elongation. Also, *N*-isobutyl-2*E*-decenamide and *N*-isobutyldecanamide were found to stimulate root hair elongation at concentrations between 10 and 8 and 10 and 7 M [27]. Similar results have been obtained using potato, tobacco, and rice, as experimental models. Taken together these results point to alkamides as a new group of plant growth-promoting substances and open the possibility of using them as molecules for improving plant production. Besides several new and more detailed analyses of the effects of affinine on the diverse experimental models, previously described studies point out "chilcuague" as one of the Mexican plants with huge agroecological potential.

4. Discussion

4.1 The importance of the indigenous plants

Several indigenous plants are in the country since at least 20,000 years ago, and several of them helped to the initial human settlement, as an important source of food for humans and animals and in some other uses, such as medical, ludic and ornamental, or spiritual. In several cases, they were considered in a sacred level and included in several rituals and ceremonies as documented in the antique codices. Some of them persist in the diet of the people from the country.

Now, these indigenous plants have been revaluated for the nutraceutical contribution in the diet, and they are included now in a possible treatment in some metabolic or degenerative diseases, making them very attractive to the pharmaceutical industry. Some of the indigenous plant uses are involved in the restoration and maintenance of the ecological systems, since the ancient times, and they are used in unique agroecological systems. Mexico is harboring one of the most extensive collections of plants around the world, and their uses are still not well understood.

Plant	Ecology	Ecological niche	Possible uses	Possibilities of cultivation	Cultivation techniques
Tagetes erecta	Subtropical	Pollination Control of microbial, insects, and nematodes	Fungicidal Bactericidal Nematicidal Food industry	Yes	Seed production
Tagetes spp.	Subtropical	Pollination Control of microbial, insects, and nematodes	Fungicidal Bactericidal Nematicidal Food industry Essential oils production	Yes	Seed production
Cactaceae family	Arid and semiarid	Erosion control and soil restauration	Ornamental Soil restoration Erosion control Food Industry Ludic	Yes	Seed production Stems Vegetative propagation Tissue culture
Heliopsis longipes	Tropical and subtropical	Unique, control of microbial, insects, and nematodes	Bactericidal Fungicidal Food industry Pharmacological Growth regulator	Yes	Seed production Vegetative propagation Tissue culture

Table 4.
Ecological characteristics of indigenous Mexican plants.

4.2 Modern uses for indigenous plants

In order to identify the importance of the plants, several scientific efforts are in the course trying to determine the biological activity of the plants. **Table 4** includes a compilation of characteristics of the plants described in the chapter. As seen several of the uses are focused on the agronomic characteristics as substitutes of the agrochemical compounds for the control of bacteria, fungi, insect, and nematodes making them very attractive for biological control in agriculture.

The biological functions are based on the secondary metabolite content; they make the plants very attractive for food industry, coloring, and tasting and even give structure to the new food products.

In this new era, the medicinal and ludic uses are ligated to avoid the mental sickness in the new societies. The ancient plants were used since the past with these proposals, with very good results, and now there is a promising use in the treatment of mental disease difficult to control in other way.

4.3 Endangered species

One of the problems in the endogenous species is the extraction of plantlets and adult plants from their habitat and being stolen for commercial purposes. It makes the plants go to a critical stage of surveillance. In several cases, the discovery of new uses has carried out an exhausted exploiting of plants without a program for reproduction and conservation of the plant species. Another situation is related to the genomic content in these species, complex polyploids with huge sizes of genome, and although they are important as ancient plants, no sequencing problems are under way.

5. Conclusions

Mexico is the center of origin for several plants, because of its privileged geographical position that includes the orography and the microclimates. The indigenous plants have been used since centuries by the inhabitants. There are several antique codices that describe the origin and uses of almost all the plants, that include ludic, ceremonial, ornamental, medicinal, and agroecological. Several Mexican plants are factories of secondary metabolites with biocide activity, such as microbial (bactericidal, fungicidal, and nematicidal) and insecticidal, with an important use in agriculture. Some other plants could help in the preservation of the soil structure, collecting water and keeping the soil in the optimal conditions for agriculture. Although everyday there is more information regarding the native plants, more knowledge is still needed in order to preserve these valuable plants.

Acknowledgements

The authors thank the Universidad De la Salle Bajío Campus Campestre for financial support.

Author details

Arellano-Perusquía Abraham, Bañuelos-Hernández Bernardo, Urquieta-Ramírez Luz, Albarrán-Tamayo Froylán, Aguilar-Ruiz Carlos Agustín, González-Márquez Marco Antonio, Junquera-Martínez Sergio and Cruz-Hernández Andrés*
Universidad De La Salle Bajío, Campus Campestre, Col lomas del Campestre, León, Guanajuato, México

*Address all correspondence to: andrex1998@hotmail.com

References

[1] Tilman D, Reich P, Knops J. Biodiversity and ecosystem stability in a decadelong grassland experiment. Nature. 2006;**441**:629-632

[2] Hughes JB, Daily GC, Ehrlich PR. Population diversity: Its extent and extinction. Science. 1997;**278**:689-669

[3] Mittermeier R, Goettsch C. La importancia de la diversidad biológica de México, México ante los retos de la biodiversidad. México: CONABIO; 1992. 14 pp

[4] Soule J. Systematics of Tagetes (*Asteraceae-Tageteae*) [Doctoral Thesis]. Austin, USA: The University of Texas; 1993

[5] Turner B. The Comps of Mexico-A systematic account of the family Asteraceae. Phytologia Memoirs. 1996;**6**: 1-93

[6] Serrato M. Información documental sobre el taxa *Tagetes* para dimensionar su centro de origen y diversidad genética en México. México: CONABIO; 2010. p. 63

[7] Comisión Nacional para el Conocimiento y Uso de la Biodiversidad. México: CONABIO; 2012

[8] Alexander S, Waldenmaier C. Suppression of *Pratylenchus penetrans* populations in potato and tomato using African marigolds. Journal of Nematology. 2002;**34**:130-134

[9] Ireri L, Kongoro J, Ngure P, Mutai C, Langat B, Tonui W, et al. The potential of the extracts of *Tagetes minuta* Linnaeus (Asteraceae), *Acalypha fruticosa* Forssk (Euphorbiaceae) and *Tarchonanthus camphoratus* L. (Compositae) against *Phlebotomus duboscqi* Neveu Lemaire (Diptera: Psychodidae), the vector for leishmania major yakimoff and schokhor. J Vector Borne. 2010;**47**:168-174

[10] Jacobs J, Aroo R, De Koning E, Klunder A, Croes A, Wullems G. Isolation and characterization of mutants of thiophene synthesis in *Tagetes erecta*. Plant Physiology. 1995; **107**:807-814

[11] Ploeg A. Greenhouse studies on the effect of marigolds (*Tagetes spp.*) on four *Meloidogyne* species. Journal of Nematology. 1999;**31**:62-69

[12] Ploeg A, Maris P. Effect of temperature on suppression of *Meloidogyne incognita* by *Tagetes* cultivars. Supplement to the Journal of Nematology. 1999;**31**:709-714

[13] Sanches F, Figueira G, Mendez A, Rodrigues B, Camargo M, Szabó M, et al. Acaricidal activity of ethanolic extract from aerial parts of *Tagetes patula L* (Asteraceae) against larvae and engorged adult females of *Rhipicephalus sanguineus* (Latreille, 1806). Parasites & Vectors. 2012;**5**. aprox. 11 pp

[14] Peña M, Cortés A, Avila E. Evaluación de tres niveles de pigmento de flor de cempasúchil (*Tagetes erecta*) sobre la pigmentación de la piel en pollos de engorda. Téc Pecu Mex. 2004; **42**:105-111

[15] Bravo-Hollis H. Las Cactáceas de México. Vol. 1. México, D.F.: Edit. UNAM; 1978

[16] Bravo-Hollis H, Sánchez-Mejorada H. Las cactáceas de México. Vol. II–III. México, D. F: Universidad Nacional Autónoma de México; 1991. 404 p

[17] Hernández MH, Godínez AH. Contribución al conocimiento de las cactáceas mexicanas amenazadas. Act. Bot. Mex. 1994;**26**:33-52

[18] Scheinvar L. Flora Cactológica del Estado de Querétaro. México, D. F.: Diversidad y Riqueza. Fondo de Cultura Económica; 2004. pp. 57-58

[19] Guzmán MS. Biodiversidad genética y caracterización nutrimental del frijol (*Phaseolus vulgaris* L.) y su potencial para mejorar el frijol cultivado [Tesis de Doctorado]. Irapuato, Guanajuato: CINVESTAV; 2001. 168 pp

[20] Ximenez F. Quatro Libros de la Naturaleza, y Animales que Aftan Recevidos en el Vfo de Medicina de la Nueva España. Mexico City, Mexico: Viuda de Diego López Davalas; 1615

[21] Little EL. Chilcuague (*Heliopsis longipes*) an insecticidal plant. Boletín de la Sociedad Botánica de México. 1948;7:23

[22] Christensen L, Lam J. Acetylenes and related compounds in Heliantheae. Phytochemistry. 1991;**30**:11-49

[23] Hofer O, Greger H, Robien W, Werner A. 13C NMR and 1H lanthanide induced shifts of naturally occurring alkamides with cyclic amide moieties: Amides from *Achillea falcata*. Tetrahedron Letters. 1986;**42**:2707-2716

[24] Parmar V, Jain S, Bisht K, Jain R, Taneja P, Jha A, et al. Phytochemistry of the genus *Piper*. Phytochemistry. 1997; **46**:597-673

[25] Rios-Chavez P, Ramírez-Chávez E, Armenta-Salinas C, Molina-Torres J. *Acmella radicans* var. radicans: *In vitro* culture establishment and alkamide content. *In Vitro Cellular & Developmental Biology. Plant*. 2003;**39**: 37-41

[26] Molina-Torres J, Salgado-Garciglia R, Ramírez-Chávez E, del Rio R. Purely oleofinic alkamides in *Heliopsis longipes* and *Acmella* (*Spilanthes*) *oppositifolia*. Biochemical Systematics and Ecology. 1996;**24**:43-47

[27] Ramírez-Chávez E, López-Bucio J, Herrera-Estrella L, Molina-Torres J. Alkamides Isolated from plants promote growth and alter root development in Arabidopsis. Plant Physiology. 2004; **134**:1058-1068

Chapter 6

Instrumental Methods for Detection of Lipophilic Marine Toxins in Endemic Species from Pacific Austral Fjords

Carlos García, Javiera Oyaneder-Terrazas and Héctor R. Contreras

Abstract

Lipophilic marine toxins (LMTs) are a group of marine toxins which in recent years have been consistently identified in the vast majority of shellfish worldwide. One of their main characteristics is having a latitudinal variability and an assimilation/retention specific for each species. LMTs consist of four important groups: okadaic acid group (OA-group), pectenotoxin group (PTX-group), azaspiracid group (AZA-group) and yessotoxin group (YTX-group). These groups have different chemical structures, which has generated an important challenge to establish analytical techniques to identify all toxic analogues from the same toxic matrix. Likewise, in the aquatic environment, shellfish represent the best bio-indicator model that allows for the establishment of levels of toxicities related to LMTs. In this chapter, the evolution for detection of LMTs from mouse bioassay (MBA), enzymatic assays (PP2a), and analytical techniques, such as liquid chromatography tandem-mass spectrometry (LC-MS/MS), are described. These analytical advances have allowed us to determine and identify the characteristic profiles of LMTs produced by marine microalgae, including the prevalence and biotransformation of LMTs in the different endemic species. It is worth mentioning that these techniques have favoured the updating of numerous sanitary standards and the definition of the most appropriate technique for the detection of LMTs in shellfish and endemic species.

Keywords: lipophilic marine toxins, AZA-group, OA-group, PTX-group, YTX-group, shellfish, mouse bioassay, protein phosphatase, liquid chromatographic, risk assessment

1. Introduction

Harmful algal blooms (HABs) consist of a group of cells (dinoflagellates, cyanobacteria and diatoms) that, under the interaction of multiple environmental factors such as luminous intensity, temperature, nutrients and salinity, among others, can increase their cell density when compared to the base population of cells present in the sea, lakes and rivers. At the same time, these HABs are associated with the

IntechOpen

capacity to produce powerful toxins that may affect the marine life of fish, marine larvae, mammals and people [1, 2].

Over the past few years, HABs have been showing a greater frequency and intensity [3–5]. Among the factors that have been proposed to be related to this global increase in HABs (**Figure 1**), the following are highlighted: eutrophication of coastal waters as a result of increased aquaculture and/or from fertilisers derived from agriculture, activities linked to urbanisation, climate change, transport of ballast water from ships and/or the transfer of shellfish populations from already contaminated areas [6, 7].

HABs associated with the number of toxin-producing cells can vary considerably during the year. Periods of exponential growth may occur, which is perceived as a bloom, during changes in weather conditions, water temperature, turbulence (waves), water salinity, and concentration of dissolved nutrients [8–10]. These environmental factors can directly affect the variability and concentration of toxins produced by each of the species involved in a bloom [7, 9, 11].

Due to the above, it is possible to establish that environmental conditions may favour the algal blooms of more than one toxin-producing microalgae, thus generating simultaneous harmful processes in the shellfish and lake ecosystems. Therefore, aquatic organisms can assimilate more than one toxic variety in their tissues, expanding the toxic effects to the living organisms in the system [10, 12, 13].

The main organisms affected by HABs are aquatic filter-feeder organisms, such as bivalves, which are characterised by having a high water filtration capacity (20 L h^{-1}) which allows them to accumulate high levels of nutrients, in addition to the accumulation of toxic phytoplankton in their tissues. All these processes are variable and dependent on endemic species in different areas of the planet [2, 14].

Several studies have provided strong scientific evidence stating that the different endemic species of bivalves identified worldwide are capable of exercising natural control of phytoplankton in coastal waters. This control process is characterised by two stages: (a) Filtration stage: this involves the accumulation of harmful dinoflagellates and their toxins in the digestive glands of shellfish (hepatopancreas); (b) Distribution stage: this involves the distribution of toxins to non-visceral tissues of shellfish, such as the mantle, gills, foot and adductor muscle [15–17]. This

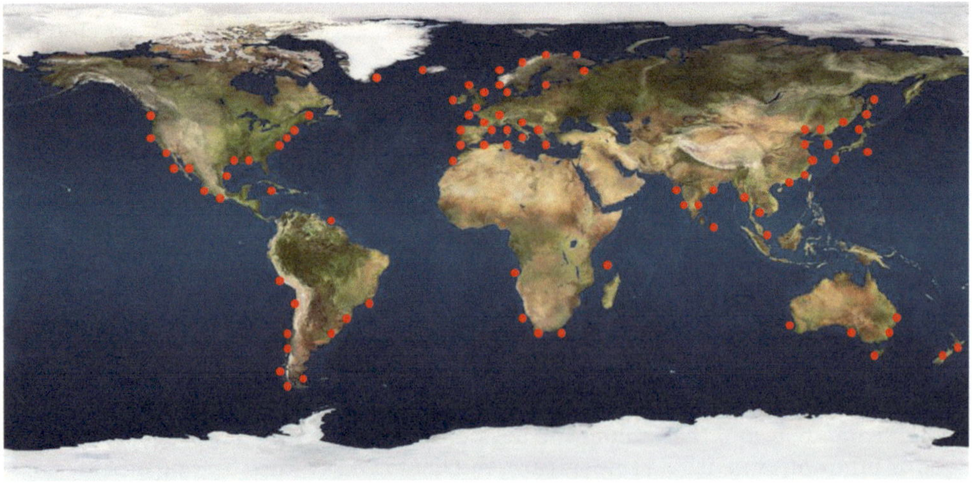

Figure 1.
Geographic distribution of harmful algal blooms associated with the identification of lipophilic marine toxins [1, 3, 5, 14].

distribution to the non-visceral tissues favours the variation of the toxic profiles assimilated in the first stage, through biotransformation pathways that involve enzymatic and non-enzymatic processes [18, 19]. These processes are enhanced by the transfer of toxins through the trophic chain, where an accumulation of toxins with very different profiles is produced in each of the marine organisms involved in the process (zooplankton and whales) [15, 20, 21]. The above mentioned stages are characterised by not causing any apparent damage to the structures and cellular composition of shellfish and bivalves, allowing these toxic compounds to remain in the digestive glands for prolonged periods (months) [22].

The prediction of HABs is very difficult, as there are a number of factors to be considered, such as physical parameters (meteorological and weather parameters, temperature, wind and light conditions, as well as hydrography), chemical parameters (nutrient variability, eutrophication, oxygen availability, anthropogenic pollution and ocean acidity) and biological parameters (evolution of algal communities, grazing and interaction of parasitic microorganisms or viruses) [7].

However, the knowledge of background information related to the temporal and geographical distribution of HAB-producing species is important for the understanding of the problems at a global level [23].

Consumption of seafood products, mainly filter-feeding shellfish contaminated with high amounts of phycotoxins in its visceral (digestive glands) and non-visceral (mantle, gills and foot) tissues, tends to produce severe intoxication when consumed by humans [24]. The number of people intoxicated with these phycotoxins worldwide has reached an average of 60,000 cases per year. This, in turn, produces a large impact on the local economy due to its negative effects on tourism, recreation, and miticulture and aquaculture industries [25, 26]. In Europe, the losses estimated every year in tourism as a result of HABs are approximately € 700 million and about € 116 million in miticulture [1, 4, 27]. Likewise, in order to prevent poisoning caused by the consumption of shellfish or hydrobiological organisms contaminated with phycotoxins, international entities have developed regulations, legislation and follow-up programs for these HABs [28–31].

1.1 Lipophilic marine toxins

Marine toxins are categorised into different groups, which are characterised by different chemical structures and different mechanisms of toxicity [32]. These groups can be produced by the same or by different species of marine microalgae [31].

Within these groups of toxins, lipophilic marine toxins (LMTs), formed by the following groups, are noted: okadaic acid- (OA), pectenotoxin- (PTX), yessotoxin- (YTX) and azaspiracid- (AZA) group.

1.1.1 Okadaic acid group (OA-group)

The OA-group consists of chemical structures formed by transfused polyethers that at its polar head have a carboxylic acid. The toxins making up this group are called: okadaic acid (OA), dinophysistoxin-1 (DTX-1, 35-R-methyl-OA), dinophysistoxin-2 (DTX-2, 31-demethyl-35-S-methyl-OA) and dinophysistoxin-3 (DTX-3, acyl derivatives of OA, DTX-1 and DTX-2) (**Figure 2**). The OA-group stand out for being potent inhibitors of 1, 2A and 3 serine/threonine protein phosphatases (PP1, PP2a and PP3) [33, 34].

The consumption of shellfish contaminated with this group of toxins produces the medical condition called diarrhoeic shellfish poisoning (DSP), which is characterised by producing symptoms such as diarrhoea, vomiting, nausea and abdominal pain [4, 31].

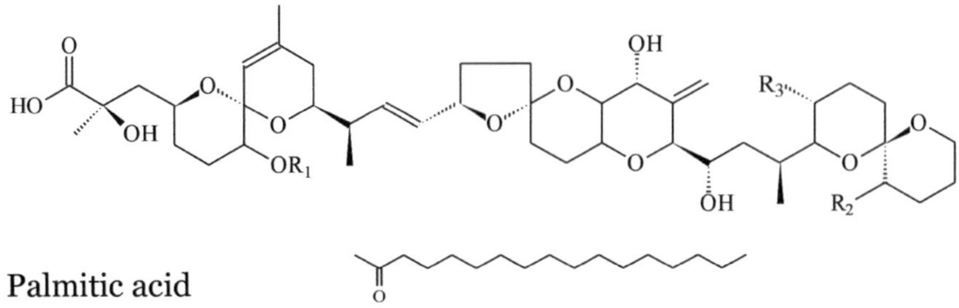

Name		R^1	R^2	R^3
Okadaic acid	(OA)	H	H	CH_3
Dinophysistoxin-1	(DTX-1)	H	CH_3	CH_3
Dinophysistoxin-2	(DTX-2)	H	CH_3	H
Dinophysistoxin-3	(DTX-3)	Acyl	CH_3	CH_3

Figure 2.
Chemical structures of okadaic acid group and analogues [33, 34].

At the beginning of the 1990s, this group of toxins was also formed by the PTX, YTX and AZA groups, since the process of extraction and evaluation of toxins in bivalves did not differentiate the toxic groups involved in the HABs processes. Thus, the toxic symptoms detected in people were associated with diarrhoeic shellfish poisoning (DSP). Subsequently, the use of analytical processes established that this group was made up of multiple toxins, which had differences from a chemical and toxicological point of view, so that each type of toxin was excluded from the OA group and classified as PTX, YTX and AZA groups [28, 35, 36].

OA-group is present in both planktonic and epibenthic dinoflagellates of the genera Dinophysis and Prorocentrum, where the species *Dinophysis acuminata*, *Dinophysis acuta*, *Dinophysis caudata*, *Dinophysis fortii*, *Dinophysis miles*, *Dinophysis ovum*, *Dinophysis sacculus*, *Dinophysis rotundata*, *Dinophysis tripos* and *Prorocentrum lima*, *Prorocentrum belizeanum*, *Prorocentrum concavum* and *Prorocentrum hoffmannianum* (**Table 1**) are highlighted [4]. The densities of the dinoflagellates associated with OA-group correspond to $<10^4$ L^{-1} cells, with a relative abundance of $<10^3$ L^{-1} cells [37].

The variability of the toxic profile identified in OA-group producing dinoflagellates depends, to a large extent, on the species involved and on the global distribution in which the blooms are identified (**Figure 1**). The variability of the toxin content detected in cells of *Dinophysis acuta* ranges between 0 and 40 OA pg. $cell^{-1}$, 0–0.02 DTX-1 pg. $cell^{-1}$ and 0.3–0.6 DTX-2 pg. $cell^{-1}$, while for *Dinophysis acuminata*, it is 0–160 OA pg. $cell^{-1}$, 0–7.8 DTX-1 pg. $cell^{-1}$ and 0–169 DTX-2 pg. $cell^{-1}$ [38].

At present, data on the chronic effects of OA in animals or humans have been insufficient to determine the Tolerable Daily Intake (TDI). However, a Lowest Observed Adverse Effect Level (LOAEL) corresponding to 50 μg OA equivalent per person has been established, equivalent to ≈0.8 μg OA kg^{-1} of body weight for adults, and a No Observed Adverse Effect Level (NOAEL) corresponding to ≈0.3 μg OA equivalents kg^{-1} b.w. [14].

Considering the lipophilicity of the toxins that make up this group, they are easily accumulated in the tissues of the filter-feeding marine organisms that feed on HABs, spreading quickly to their predators in the food chain [13]. Moreover, depending on the species of molluscs contaminated with the OA-group and their natural clearance, the OA-group toxins tend to be chemically modified in the visceral tissues of the molluscs. In this way, OA, DTX-1 and DTX-2 can be esterified by fatty acids of variable length (C7-C22) in the hydroxyl group (–OH) present in carbon 7 of the structure of toxins (**Figure 2**), palmitic acid (C16:0)

Species	Toxins produced
Dinophysis sacculus	OA
Dinophysis ovum	OA
Dinophysis mitra	DTX-1
Dinophysis tripos	DTX-1
Dinophysis miles	OA, DTX-1
Dinophysis rotundata	OA; PTX-2
Dinophysis infundibulus	PTX-2
Dinophysis caudata	OA, DTX-1, PTX-2
Dinophysis fortii	OA, DTX-1, PTX-3
Dinophysis norvegica	OA, DTX-1, PTX-2, PTX-12
Dinophysis acuminata	OA, DTX-1; PTX-2; PTX-12
Dinophysis acuta	OA, DTX-1, DTX-2, PTX-2, PTX-2sa
	7-epi-PTX-2sa, PTX-12
Prorocentrum concavum	OA
Prorocentrum hoffmannianum	OA
Prorocentrum maculosum	OA
Prorocentrum mexicanum	OA
Prorocentrum belizeanum	OA, DTX-1
Prorocentrum faustiae	OA, DTX-1
Prorocentrum lima	OA, DTX-1
Protoperidinium divergens	OA, DTX-1
Protoperidinium depressum	OA, DTX-1
Protoceratium reticulatum	YTX
Gonyaulax spinifera	YTX
Lingulodynium poliedrum	YTX, homo-YTX, 45-OH-homo-YTX
Protoperidinium crassipes	AZA
Echinochlathria sp.	AZA-2
Azadinium spinosum	AZA-1, -2, -3

Table 1.
List of dinoflagellate species identified as producer species of lipophilic marine toxins [4, 28, 37, 55, 64, 65, 79, 81].

being the most prevalent that is identified in this process [39, 40]. This esterification results in the toxic structures being named acyl derivatives. However, in their first detections, acylation was always associated with DTX-1, acquiring the name of dinophysistoxin-3 (DTX-3) [41]. In Chile, palmitic acid accounts for 90% of the total fatty acids linked to the esterification of toxins in the OA-group, in which 7-O-palmitoyl-dinophysistoxin-1 is the most prevalent toxin detected in endemic bivalves on the coast of Chile [39, 42]. Although the feasibility of other forms of fatty acid derivatives of varying lengths or with different unsaturations may produce acyl derivatives of the OA-group, such as C14:0, C16:1, C16:0, C18:1 and C18:0 [43–45]. All these 7-O-acyl-OA-group esterified analogues have not been detected in any single-cell isolates of *Dinophysis* sp., so detection is only associated with shellfish [46].

The differences observed between the proportions of esterified forms and free forms of OA could arise from the genetic differences in bivalves (endemic species), since the OA esterification is considered to be an enzymatic mechanism associated with the detoxification of bivalves [47]. From a toxic point of view, it has been established that the toxicity of endemic species may depend on the variability in the lipid content present in the digestive glands, which would favour the retention of the toxins, thus explaining its higher toxicity of compartmentalisation if compared to non-visceral tissues, such as the mantle or adductor muscle [19, 48, 49].

The symptoms caused by intoxication associated with the consumption of molluscs contaminated with the OA-group are characterised as starting between 1 and 5 h after the ingestion of contaminated molluscs, symptoms that tend to be reversed 3 days after the toxic symptoms are initiated [50]. The minimum doses of OA and DTX-1 required to produce toxic symptoms in humans have been estimated to be 40 and 36 μg respectively for a 60 kg person [51]. To date, no deaths of people associated with intoxications produced by the OA-group have been recorded. However, it has been clearly established that the OA-group is a potent tumour promoter in animals [52], thus being associated with the risk of gastric cancer among regular consumers of contaminated shellfish with toxins of the OA-group [53, 54]. Therefore, poisonings associated with the OA-group are a latent problem for both public health and the seafood industry [1, 4, 27].

1.1.2 Pectenotoxin group (PTX-group)

Toxins making up the pectenotoxin group (PTXs) correspond to a family of toxins of polyether macrolides (**Figure 3**) produced by the same species that produce the toxic forms of the OA-group, such as *Dinophysis fortii*, *Dinophysis acuta*, *Dinophysis acuminata*, *Dinophysis caudata* and *Dinophysis norvegica*, which are further detected in heterotrophic dinoflagellates such as *Protoperidinium divergens* and *Protoperidinium depressum* (**Table 1**) [28, 55]. The origin of the name of this group comes from the crustacean *Patinopecten yessoensis* in which it was detected for the first time [33].

For years, the PTX-group was included in the OA-group, since the bioassay tests detected the overall toxicity of both groups. Currently, the PTX-group is classified as a separate group, based on the biochemical effects on which they act, as this group does not inhibit protein phosphatase 2a (PP2a) [14, 56]. To date, no cases of intoxication due to consumption of molluscs contaminated with the PTX-group have been reported [57].

This group of toxins has been identified in countries such as Ireland, Croatia, New Zealand, Portugal, Norway, Japan, Argentina and Chile, showing a direct relationship with the areas which have previously been associated with the presence of the OA-group (**Figure 1**) [28, 40, 58, 59].

The toxic profiles detected in dinoflagellates and shellfish tend to be different, establishing that a metabolic transformation occurs in the bivalves after the filtration/assimilation of toxic algae [46]. Filter-feeding bivalves, when accumulating this type of toxin (PTX-2) in their digestive glands (hepatopancreas), allow for their biotransformation. These biotransformed analogues have been assigned to the seco acid (sa) nomenclature, highlighting the identification in bivalves of pectenotoxin-2 seco acid (PTX-2sa) and 7-epi-pectenotoxins-2-seco-acid (7-epi-PTX-2sa) analogues [28, 43]. 7-epi-PTX-2sa, is the result of the interconversion of PTX-2sa to a thermodynamically more stable analogue. In this way, the conversion of PTX-2 to seco-acid forms could be considered as a protective effect of detoxification by molluscs [41].

There are about 15 analogues identified and associated with the PTX-group, in which the esterified forms of PTX-2sa were the last identified analogues, where

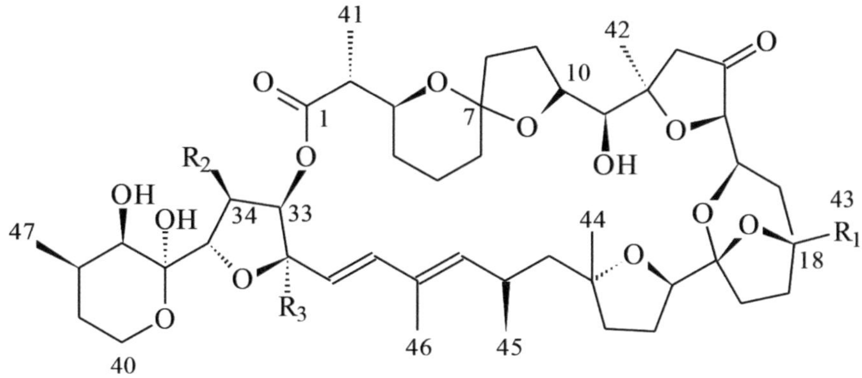

Name	R[1]	R[2]	R[3]	C-7
Pectenotoxin-1 (PTX-1)	CH_3OH	H	H	R
Pectenotoxin-2 (PTX-2)	CH_3	H	H	R
Pectenotoxin-2b (PTX-2b)	CH_3	H	H	S
Pectenotoxin-3 (PTX-3)	CHO	H	H	R
Pectenotoxin-4 (PTX-4)	CH_2OH	H	H	S
Pectenotoxin-6 (PTX-6)	COOH	H	H	R
Pectenotoxin-7 (PTX-7)	COOH	H	H	S
Pectenotoxin-11 (PTX-11)	CH_3	OH	H	R
Pectenotoxin-11b (PTX-11b)	CH_3	OH	H	S
Pectenotoxin-13 (PTX-13)	CH_3	H	OH	R

Figure 3.
Chemical structure of pectenotoxin group and its chemical analogues identified in dinoflagellates and bivalve molluscs [28, 40].

37-O-acyl ester-PTX-2-sa, 11-O-acyl ester-PTX-2sa and 33-O-acyl ester-PTX-2sa are noted. All of these analogues were identified in filter-feeding bivalves [60–62].

The available toxicological information on the PTX-group is insufficient at present due to the lack of toxic analogues required to carry out biological studies of interest. However, it has been determined that the lethal intraperitoneal dose of PTX-2 corresponds to 260 μg kg^{-1}. In addition, the results of the toxicity obtained from intraperitoneal (*i.p.*) administration and oral ingestion of PTX in mouse have established that toxicities are comparable, do not produce these symptoms of diarrhoea and are only limited to generate histopathological changes in the liver, stomach and intestine [60]. Furthermore, it has been determined that PTX-2 may potentially be cytotoxic on lung, colon and kidney cell lines. Toxic effects are not extrapolated to analogues such as PTX-2-sa and 7-epi-PTX2-sa, evidencing the importance of the initial structure of the toxin that generates the cytotoxic effects [63].

1.1.3 Yessotoxin group (YTX-group)

Yessotoxin group (YTX-group) corresponds to a group constituted by sulphated polyethers (**Figure 4**), and whose analogues were first identified in the oyster *Patinopecten yessoensis*. Yessotoxin (YTX) is produced by the marine phytoplanktonic microalgae *Protoceratium reticulatum* (Gonyaulax grindley) [64], *Lingulodinium polyedrum* (*Gonyaulax polyedra*) [35, 65] and *Gonyaulax spinifera* (**Table 1**) [35, 66, 67].

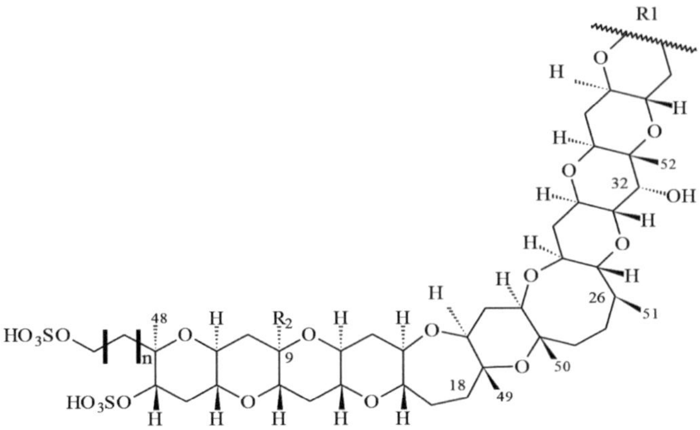

Name	R¹	R²	n
Yessotoxin (YTX)	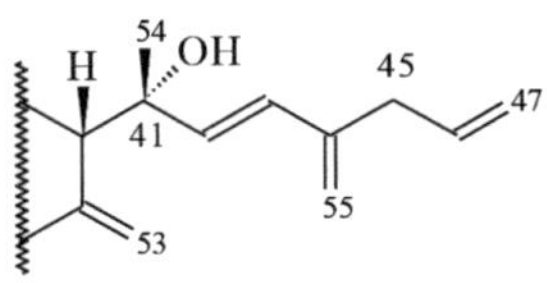	H	1
45-Hydroxy-YTX (45-OH-YTX)	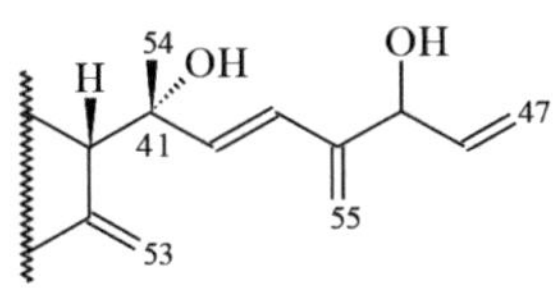	H	1
Homo-YTX (1a-Homo-YTX)	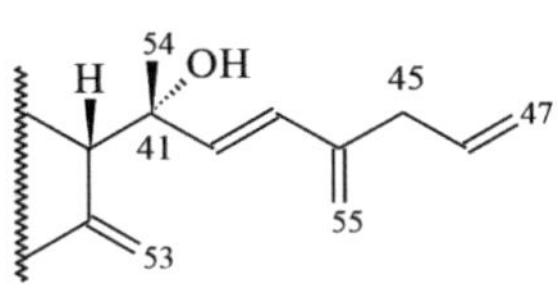	H	2

Figure 4.
Chemical structure of yessotoxin-group and their chemical analogues identified in dinoflagellates and bivalve molluscs [64, 66].

At present, this group of toxins have been identified in different countries worldwide such as New Zealand, Italy, Spain, Norway, Russia, Canada, United Kingdom, Japan, Argentina and Chile (**Figure 1**) [35, 68–70].

Historically, the YTX-group was included in the OA-group, however, this group of toxins do not produce diarrhoea in the mouse bioassay, nor do they produce the inhibition of the protein phosphatase 2a (PP2a) [71, 72]. Although the precise mode of action of the YTX-group is unknown, it has been classified as a potent activator of phosphodiesterase [73].

To date, more than 100 natural YTX analogues have been identified and characterised using nuclear magnetic resonance (NMR) and liquid chromatography coupled with mass spectrometry (LC-MS). Some of the identified analogues are directly related to the producer dinoflagellate, such as norYTX, 41-keto-YTX and 41a-homo-YTX, while other analogues come from processes of biotransformation exerted in the digestive glands of different marine species worldwide that have been evaluated and that involve chemical oxidation pathways such as hydroxylations, carboxylations, desulfations, methylations and amidations [35, 61]. Thus, analogues such as 45-hydroxy-YTX, homoYTX, 45-hydroxyhomo-YTX, carboxy-YTX, carboxyhomo-YTX, 41a-homoYTX, 45-hydroxycarboxy-YTX, and 1-desulphocarboxyhomo-YTX are the direct result of biotransformation processes that occur in shellfish [28, 46, 74].

The HABs associated with the dinoflagellates producers of the YTX-group are characterised by reaching a cell density of 10^3 L^{-1} cells, where the production of YTXs in dinoflagellates is on average ≈34 pg. $cell^{-1}$ (0–74 pg. $cell^{-1}$), where homo-YTX and YTX are the main toxins forming the profile, whose concentrations are variable and dependent on the areas where blooms have been detected [35].

Symptoms caused by intoxication with the YTX-group in humans are unknown, because no human poisonings associated with this group have been reported [28]. However, the toxic evaluation of YTX in bioassays has determined that an intraperitoneal injection at concentrations of approximately 150 μg kg^{-1} causes difficulty in mobilisation, dyspnoea, jumps, tremors and cramps, with all symptoms starting 4 h after the injection is given [67, 71, 72].

It is noteworthy that YTX oral administration in mouse did not induce any significant difference in haematological and clinical chemistry parameters, including leukocyte percentages and plasma levels of alanine-aminotransferase (ALT), aspartate aminotransferase (AST), creatine phosphokinase (CPK) or lactate dehydrogenase (LDH) [67, 71, 72, 75]. Nevertheless, at an ultrastructural level, changes in the myocardium in *in vivo* studies following both oral and *i.p.* administration have been identified [71, 72, 75–77]. Thus, it has been estimated that YTX intraperitoneal administration induces cardiac damage with a potency >10 times greater than the oral route [76]. This difference could be related to the low adsorption through the gastrointestinal tract and/or biotransformation of toxins associated with the YTX-group [67].

In addition, toxicological data obtained from the evaluation with the YTX and homo-YTX analogues have shown that they have approximately the same toxicity [64], while the other analogues have lower toxicity, especially OH-YTX and carboxy-YTX derivatives because they are ≈5 times less toxic than YTX, while other derivatives associated with the YTX group, such as trihydroxylated amides (41-a-homo-YTX and 1,3-enone isomer of heptanor-41-oxo YTX) have shown no toxicity through intraperitoneal injections in mice at levels >5000 μg kg^{-1} body weight [7, 60, 61].

Naturally, the main vectors of the YTX-group correspond to bivalves (endemic species) characterised by accumulating large amounts of toxins in their digestive glands due to their high filtration capacity. Once they are assimilated and based on the chemical modifications caused by bivalves to toxic analogues of YTX, they can be spread to other non-visceral tissues such as mantle, gills, foot and adductor muscle [62, 74].

The European Food Safety Authority (EFSA) has established that toxic effects may even occur at concentrations below 3.75 mg YTX equivalents kg^{-1} shellfish, which is a limit established as an international sanitary standard for marketing products [14, 36, 78].

1.1.4 Azaspiracid group (AZA-group)

AZA-group corresponds to toxins produced by toxic dinoflagellates, *Protoperidinium crassipes*, *Azadinium spinosum* and the sponge, *Echinochlathria* sp. (**Table 1**) [79–81]. However, in some species it has not been possible to establish a direct relationship between the high toxic levels detected in some species of endemic bivalves and the cell densities of the dinoflagellates producing this group of toxins [82].

The AZA-group has been identified worldwide in molluscs analysed in Ireland, United Kingdom, Norway, France, Portugal, North Africa (Morocco), Chile and USA (**Figure 1**) [12, 43, 44, 83].

To date, ≈24 toxic analogues belonging to the AZA-group have been identified. The main toxic analogues detected are: azaspiracid-1 (AZA-1), azaspiracid-2 (AZA-2) and azaspiracid-3 (AZA-3) (**Figure 5**) [84, 85].

The mechanisms of action of the AZA-group have not been clearly determined, however studies *in vitro* in mammalian cells show that they produce alterations in the structure of the cytoskeleton of cells and the E-cadherin system, the latter being responsible for the interaction between cells [86, 87]. These experimental results could explain the symptoms associated with human intoxications such as gastrointestinal disorders, abdominal pain and diarrhoea [1]. Although the data obtained from seafood extracts have shown that AZAs do not produce diarrhoea [88].

The intraperitoneal lethal dose determined in mice is directly related to the toxic analogue detected in shellfish extracts. In this way, AZA-1 has a lethal dose of 200 μg kg^{-1}, while AZA-2 and AZA-3 are significantly more toxic, with lethal doses of 110 and 140 μg kg^{-1}, respectively [28]. In addition, it has been established that oral toxicity in mice with AZA-1 at corresponding doses of 900 μg kg^{-1} produces significant damage to the small intestine, while doses of 500 μg kg^{-1} only produce liver damage, which is characterised by an increase in the volume (≈38%) [89, 90].

In addition, toxicological studies have established that the lowest observed adverse effect level (LOAEL) for AZAs corresponds to concentrations between ≈23 and ≈86 μg AZAs kg^{-1} per person, even though it is estimated that at levels of ≈80 μg AZA-equivalents kg^{-1} molluscs would not produce symptoms due to AZA-group intoxication. However, it has been recorded that when ≈400 g of shellfish are consumed, doses corresponding to ≈30 μg AZA-equivalents kg^{-1} have been able to produce AZA-associated intoxication syndrome in humans. These seemingly contradictory data can be explained by the variability of toxic forms subjected to biotransformation, which produces new and different analogues during the accumulation of toxins in the digestive glands of molluscs [91].

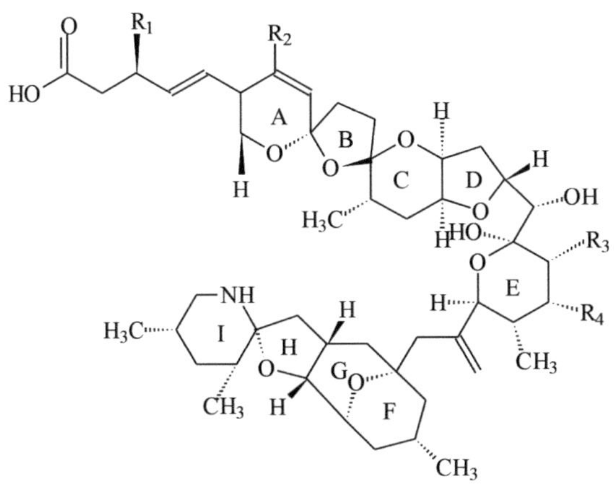

Name	R^1	R^2	R^3	R^4
Azaspiracid-1 (AZA-1)	H	H	CH_3	H
Azaspiracid-2 (AZA-2)	H	CH_3	CH_3	H
Azaspiracid-3 (AZA-3)	H	H	H	H
Azaspiracid-4 (AZA-4)	OH	H	H	H
Azaspiracid-5 (AZA-5)	H	H	H	OH
Azaspiracid-6 (AZA-6)	H	CH_3	H	H

Figure 5.
Chemical structure of azaspiracid-group and their chemical analogues identified in dinoflagellates and bivalve molluscs [79–82].

Although AZA-1 is the most abundant toxin in this group, the analogues corresponding to AZA-2, AZA-3, 8-methyl-AZA-1 and 22-demethyl-AZA-1 frequently co-occur in different types of bivalves (**Figure 5**). The formation of AZA-3 from AZA-1 in *Mytilus* sp. corresponds to a xenobiotic bioconversion, even though this detoxification pathway results in the formation of a more toxic analogue [81]. The toxicity of molluscs contaminated with AZA-group toxins tends to persist for an average period of ≈5 months post-bloom, allowing toxins to be distributed to other non-visceral tissue. In some cases, these tissues are able to reach toxic levels superior to those described for the digestive glands [12, 92].

In spite of this, the trophic transfer routes through the food web are completely unknown, although to date the AZA-group has been detected in other aquatic species such as crustaceans [43, 93].

2. Results

2.1 Methods of detection

Each group of toxins (OA, PTX, YTX and AZA) present in the different endemic species of bivalves or marine organisms is composed of many toxic analogues, so regulatory levels are represented according to the total toxicity of the analogues studied. Traditionally, regulatory limits have been evaluated using mouse bioassay (MBA), which involves intraperitoneal injection of seafood extracts [14, 91].

Nevertheless, we should consider that one single type of shellfish may contain more than one toxic group, therefore, alternative methods to the bioassays are required for the detection and proper identification of each analogue that may belong to the different described groups [94].

To identify toxic groups, there are also functional tests, which are traditionally defined as methods of detection based on the mechanism of action that each toxin group has in order to establish its quantification, which, in turn, is related to the group toxicity. These assays are usually receptors, proteins or cells [95].

Since shellfish consumption is very important from an economic standpoint, most producer and consumer countries have implemented monitoring systems to prevent these toxins from reaching consumers [94]. These monitoring systems based on detection methods are required to be highly specific, reproducible systems and must not be prone to produce either false positives or false negatives [14].

Thus, some countries differ in the selection of the method required for the identification of toxin groups and three different methods can be used: biological methods (Mouse Bioassay, MBA), biochemical methods (Protein Phosphatase-2a Inhibition Assay, PP2aIA) and chemical methods (Liquid chromatography coupled with tandem mass spectrometry, LC-MS/MS) [96]. Biological and biochemical methods only establish the presence of one group of toxins; they do not identify the toxin involved in a mixture of analogues in a process of contamination or associated with a HAB whereas chemical methods provide a profile of the quantity and variety of analogues in a contaminated sample [94].

At an international level, the identification of LMT has been regulated by establishing maximum permissible concentrations for marine organisms destined for human consumption. The maximum permissible limits are 160 μg of OA equivalents kg^{-1} shellfish meat for OA-group, PTX-group and AZA-group; and 3.75 mg of YTX equivalents kg^{-1} shellfish for YTX-group [29, 30, 78].

It is also worth noting that the OA- and PTX-groups are represented toxicologically together, but this fact is based more on the possible co-occurrence of toxins of the OA-group and PTX-group from a same species of dinoflagellates. However, it should be considered that these groups do not share the same biological action mechanisms [14, 35, 55].

This regulation allows a series of methods for the detection of these groups, noted among them, immunoassays, functional tests (inhibition of protein phosphatase) and LC-MS/MS [97].

2.1.1 Biological method: mouse bioassay (MBA)

The mouse bioassay (MBA) is a biological assay used for the determination of toxins corresponding only to the OA-group. This bioassay can only provide a response to this group of toxins, however it does not provide any information on the analogues present in contaminated shellfish [7].

To obtain results, this bioassay requires the use of at least three animals per sample, which, in the end, are euthanized regardless of the obtained toxicity result [95]. Thus, the shellfish extract is injected via intraperitoneal (*i.p.*) route to three male mice that must have a weight ranging between 16 and 20 g. A sample is considered positive (contaminated with OA-group) when at least two of the three mice die within 24 h of the injection (**Figure 6**) [95, 98].

The positive result of this bioassay on toxic samples causes diarrhoea in mouse, which is directly proportional to the concentrations of toxins present in the toxic seafood extract. The toxins detected by this MBA may even be present in the faeces of the animals when the concentrations are >700 μg kg^{-1}, with an average lethality exceeding 30% at a dose of ≥1000 μg kg^{-1} [34]. Excretion of OA-group

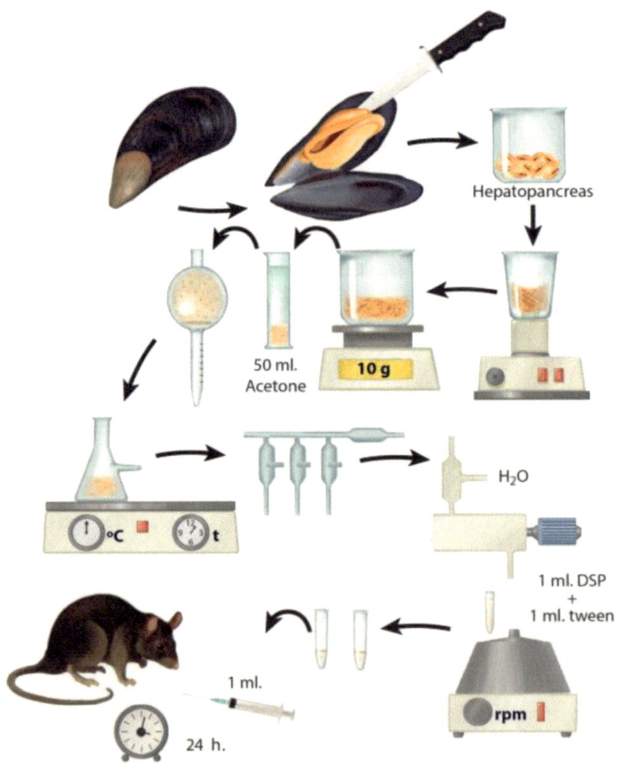

Figure 6.
Diagram of the stages required for the determination of OA-group toxins through the use of the mouse bioassay (MBA) [95, 98].

toxins through faeces indicates that a fraction of the toxin ingested by animals is not absorbed, being eliminated instantaneously, while it is possible to detect low levels in different organs (liver, pancreas, and spleen) as compared to the amount excreted in urine and faeces [34, 99].

In the absence of death (negative assays), mice may similarly develop symptoms specific to OA-group toxins which are significant indicators of a potential contamination and risk [14, 100].

However, a critical factor of the MBA is the body weight of the mouse used, which significantly influences the susceptibility of the toxins associated with the OA-group. The lethality after inoculation of OA at 24 h post-injection reaches 100% in mouse with weights of 14–15 and 16–17 g, while for groups of 19–20 g, it reaches 80% and for groups of mouse weighing between 21 and 22 g, it only reaches 50% [98].

Moreover, all toxins in the OA-group have been shown to be less toxic by oral route compared to an intraperitoneal injection route. The median lethal dose (LD_{50}) via oral route for OA is approximately 400–880 μg kg^{-1}. Specifically, LD_{50} toxicity via *i.p.* for OA averages between 192 and 225 μg kg^{-1} [101], while for DTX-1, LD_{50} toxicity via *i.p.* is ≈160 μg kg^{-1} [102] and for the DTX-3 and DTX-4 analogues, the toxicity ranges from ≈352 to ≈600 μg kg^{-1} [103]. It should be noted that the acylated forms (DTX-3, acyl-derivatives-OA-group) can show a decreased toxicity, stating that these toxic analogues can be ≈20 times less toxic than OA [101]. Additionally, preliminary data on the oral toxicity of DTX-2 samples showed that oral LD_{50} is ≈2150 μg kg^{-1} body weight [104].

These variabilities on acute toxicities of various analogues such as DTX-4 and 7-O-palmitoyl-OA are related to the bioavailability of toxins within the peritoneal cavity, which, at varying ranges, are hydrolysed under acidic or alkaline conditions or by the action of esterases or lipases within the gut allowing for the production of more toxic forms such as OA or DTX-1 [50, 101, 105].

This toxic variability between different analogues means the use of the MBA has a probability of detecting toxins between 40 and 50% associated with the OA-group in the currently established limit of 160 μg OA-equivalents kg^{-1} [7, 10, 14, 102].

Despite the above mentioned, the great advantage of the MBA is that it provides an estimate of the total toxicity of the sample [14], but, at the same time, it has multiple disadvantages:

- It requires animal facilities and expertise (bioterium).
- It cannot be easily automated due to the involvement of animal handling.
- It has a high variability in the results between laboratories, due to characteristics in the mouse used (breed, sex, age, weight, general health status, diet, stress) [98].
- It produces false-positive results due to fatty acid interferences (20,4n-6 and 20,5n-3) [106].
- It can produce false-negative results due to the toxic suppression between groups [96].
- Interfering matrices from heavy metals.
- It is not selective for the OA-group toxins [107].
- It is not a quantitative bioassay and it has a limit of detection ≈200 μg kg^{-1}.

- The *i.p.* injection of the sample is not suitable for the complete detection of the OA-group, since some analogues require hydrolysis to be detected (acyl-derivatives) [105].
- Some countries have banned it for ethical reasons.

Within these disadvantages, it has been established that MBA is able to detect toxins of the OA-, AZA- and YTX- groups. However, the result will never be able to determine the exact group responsible for it, as there is a co-occurrence of these toxins in contaminated endemic species (shellfish) [7, 12, 13, 26].

Toxic evaluations in mice related to the YTX-group have shown that they die (2/3) by injecting them with doses 0.75 mg YTX kg^{-1} body weight, while this increases (3/3) with a dose of 1 mg YTX kg^{-1} body weight. However, mice cannot be killed with oral doses of 10 mg kg^{-1} YTX kg^{-1} body weight (LD_{50} *i.p.* ≈100 μg kg^{-1}) [7]. The differences between intraperitoneal and oral toxicities of YTX would probably be related to the low YTX uptake in the gastrointestinal tract (≈0.02%) [35]; those factors are increased when considering other analogues of the YTX-group that arise from the processes of biotransformation in the tissues of marine organisms (45-hydroxy-YTX, carboxy-YTX) [7].

Regarding the toxic evaluation through MBA with the AZA-group, it has been established that this assay is capable of detecting the toxins associated with AZA-group, producing neurotoxic symptoms in mice such as: slowness, respiratory difficulties, spasms, progressive paralysis and death between 20 and 90 min after the application of the mouse bioassay. Thus, it has been determined that the minimum *i.p.* lethal dose required to cause swelling of the stomach and liver in the mouse, with a reduction in the size and weight of the thymus and spleen is ≈150 μg kg^{-1} [10].

In addition, it has been established that the different analogues of this group have different toxicities, the minimum lethal dose of AZA-2 (8-methyl-azaspiracid) and AZA-3 (22-desmethyl-azaspiracid) corresponds to ≈110 and ≈140 μg kg^{-1}, respectively, suggesting a higher potency in relation to AZA-1. However, AZA-4 and AZA-5 are less toxic, with lethal dose values of ≈470 and < 1000 μg kg^{-1}, respectively. In this way, the toxicity of these analogues through the MBA can be represented as follows: AZA-2 > AZA-3 > AZA-1 > AZA-4 > AZA-5 [10, 16].

In relation to the toxic evaluation of AZA-group by oral route, it has been shown that it does not produce death in mice at concentrations >900 μg kg^{-1} after 24 h. However, post-mortem evaluations have shown various gastrointestinal disturbances, such as accumulation of fluid from the ileum and necrosis of intestinal epithelial cells [10].

For the PTX-group, toxic data related to the MBA establish that they are highly toxic by *i.p.* injection in values averaging between ≈219 and ≈411 μg kg^{-1} [60], which leads to a positive MBA. However, PTXs appear to be of low toxicity by oral route (≈5.0 mg kg^{-1}) and, unlike OA, they do not cause diarrhoea, in addition to the fact that PTX-group is easily destroyed under basic conditions [60, 108].

In this regard, toxicity in mouse with different combined doses of LMTs by oral route has been assessed. Thus, combined oral doses of YTX (1 mg kg^{-1} day^{-1}) and OA (0.185 mg kg^{-1} day^{-1}) show no lethal effects or diarrhoea or any other symptoms of toxicity in mice, as opposed to the results obtained from the toxic evaluation of these groups individually [67].

These data are consistent with the evaluation of combined doses of yessotoxin (1.0 or 5.0 mg kg^{-1}) and AZA-1 (200 μg kg^{-1}) given to mice; those doses do not produce toxic effects on the heart or any other internal organs. It is noted that the absorption of YTX is not potentiated by the co-administration of AZA-1 [67, 109].

Furthermore, the combined evaluation of OA and AZA-1 shows that there is no increase in pathological changes when AZA-1 and OA are administered together, if compared to the administration of OA alone, where a jejunal dilatation is clearly evidenced. It should be noted the reduction of OA absorption in the internal organs when administered together with AZA-1 or vice versa, due to the competence experienced by the weak organic acids through the simple diffusion of the membranes of the gastrointestinal tract should be noted [32].

In conclusion, MBA is not the most appropriate when considering the simultaneity of the toxic bloom processes at a global level and the high capacity of biotransformation of shellfish, which would produce a high variability in the toxicity results through this bioassay.

2.1.2 Biochemical method: protein phosphatase 2a (PP2a) inhibition assay

Biochemical assays are the preferred methods for detecting LMTs in endemic species (shellfish). However, they cannot provide a quantitative measure of all toxic groups (OA-, PTX-, AZA- and YTX-group). This is the main problem of the protein phosphatase 2a (PP2a) inhibition assay, a recognised biochemical method capable of accurately detecting and quantifying only some analogues related to the OA-group [96, 107, 110, 111].

The PP2a inhibition assay (PP2aIA) is a rapid method for the detection of toxins associated with the OA-group, which is based on its functional property of inhibiting type PP2a (**Figure 7**) [112–114]. In general, this method tends to be precise, sensitive, reproducible, simple, and fast [47, 115, 116].

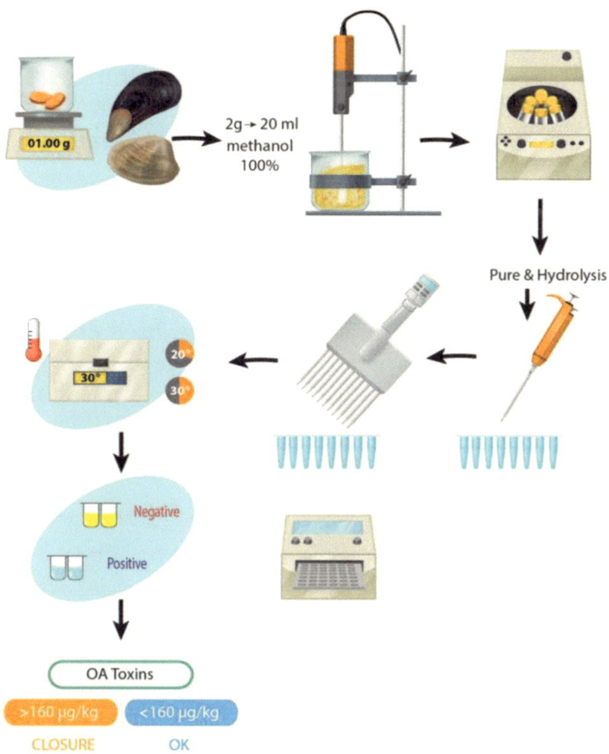

Figure 7.
Stages of the protein phosphatase 2a inhibition assay for the detection of toxins associated with OA-group toxins [107, 122].

Several methods of purification of PP2a have been proposed, including: purification of recombinant PP2a overexpressed in insect cells [114, 117, 118], PP2ac expression in mammalian cells [119], and overexpression of PP2ac in yeast [120]. However, none of these systems is capable of producing a high yield of recombinant proteins [110, 121].

The principle of the assay is based on the fact that PP2a reacts with the p-nitrophenylphosphate (p-NPP) substrate, which is colourless to produce p-nitrophenol (p-NP) by the enzymatic reaction, characterised by having a yellow colour. The toxins associated with the OA-group (OA and DTX-1) inhibit the enzyme, causing a loss of production of the p-NP; this loss is determined through the variability of the absorbance at 405 nm [107]. The amount of enzyme determines the amount of analyte (OA and DTX-1) needed for the inhibition, while the quality of the enzyme ensures the amount of product formed per time unit [122]. In addition, the lack of stability or impurity of the standards used (OA and DTX-1) directly affect the quantification [123, 124].

Regarding the OA-group, it should be noted that DTX-1 is a more potent inhibitor of PP2a than OA [125, 126], whereas DTX-2 is half as potent when compared to OA [102, 112]. The presence of esters in microalgae, such as DTX-4 is ≈500 times less active than OA [103] and acylated forms originated from the biotransformation process in shellfish, such as 7-O-palmitoyl-OA, which is a very weak inhibitor of PP2a, ≈3000 times less active than OA (**Table 2**) [101].

Additionally, the sensitivity of the enzymes (PP2a) for the evaluation of toxins associated with the OA-group may be drastically different. Thus, the choice of the enzyme and the source of origin are crucial for the proper performance of the system. Therefore, one of the main drawbacks is related to the low enzymatic stability of this assay, so, to overcome this problem, some kits have immobilised the enzyme (PP2a), which is a crucial step for the extraction of biosensors [107].

The great advantage of PP2aIA is that it provides an estimate of the total toxicity of the sample; it is a repetitive and fast assay [14]. At the same time, the PP2aIA has multiple disadvantages, among which, the following are found:

- It requires a good quality enzyme (PP2a) [123, 124].
- It requires background knowledge for the interpretation of data.
- It requires interlaboratory validation and standardised protocols.

Toxins	**PP2a**	
	Substrate	**K_i**
Okadaic acid (OA)	pNPP	30
Dinophysistoxin-1 (DTX-1)	pNPP	19
Dinophysistoxin-2 (DTX-2) (IC_{50})	pNPP	3.38
7-Deoxy-OA	pNPP	900
7-O-palmitoyl-OA	pNPP	>100 nm
7-O-palmitoyl-DTX-1	pNPP	>100 nm
Methyl-okadaate	pNPP	≫100 nm
2-oxo-Decaborxyl-OA	pNPP	≫100 nm

K_i = dissociation constant.

Table 2.
Inhibition of PP2a by OA-group [112, 122, 127].

- The matrices from marine organisms can underestimate or overestimate the results [124, 129].
- It does not provide any information on the profile of OA-group [101].
- It does not detect acyl-derivatives of OA-group directly [126, 128]
- It does not identify analogues associated with the PTX-, YTX- and AZA-groups [96, 107].

At present, it is widely accepted that the toxic effects of OA-group are caused by the inhibition of protein phosphatases. However, there is no concrete evidence supporting this hypothesis. This is because a pathway from inhibition of PP2a to the toxic effect produced by toxins in the OA-group has not been identified. The OA-group-induced toxic effects are replicated by substances which are not inhibitors of protein phosphatase (*Vibrio parahaemolyticus*), and known inhibitors of protein phosphatase (Microcystin, MC) do not exert the same toxic effects in animals just as the OA-group toxins. Therefore, this method does not have a direct relationship with the systemic effects involved in the processes of intoxication in humans, which are directly limited to the specific ability of the OA-group to inhibit PP2a [101].

However, the results of PP2aIA obtained with OA-group toxins are very well correlated with the results obtained using the MBA. The IC_{50} concentrations for DTX-2 and OA are ≈5.94 and ≈2.81 ng ml^{-1}, respectively, indicating that OA is approximately two times more toxic than DTX-2 [102].

Different origins of PP2a enzymes have allowed for comparisons of the assay, demonstrating that PP2a wild-type is significantly more sensitive to all toxins in the OA-group, relative to that obtained from a recombinant origin, which is ≈1.9-fold lower IC_{50} for OA, ≈1.7-fold lower IC_{50} for DTX-1, and ≈2.2-fold lower IC_{50} for DTX-2 [124]. Thus, the PP2aIA can be considered to detect toxins associated with toxins of the OA-group since its limit of detection (LOD) and limit of quantification (LOQ) are <160 μg OA equivalents kg^{-1} (international maximum limit). Nevertheless, its optimal range of toxic detection is very narrow (between ≈56 and ≈96 μg OA kg^{-1} shellfish) due to the fact that matrices and pigments of bivalves may affect the interpretation of results [14, 123]. It should be considered that samples with DTX-2 only are not able to adequately inhibit PP2a. In this way, and comparing the different types of enzymes according to their origin, it has been generally established that recombinant PP2a tend to underestimate the equivalent contents of OA, whereas the tendency of PP2a wild-type is to overestimate the toxicities [124].

Moreover, the process to establish the total toxicity of a sample of shellfish exposed to HABs associated with OA-group through PP2aIA necessarily involves the hydrolysis of the samples from shellfish, in order to establish the ranges of acyl-derivatives, which are only detected in shellfish and which are also notable for not producing a very low inhibition of the different types of PP2a [126, 128]. Thus, the acylated analogues must be transformed to their base analogue, which could be OA, DTX-1 or DTX-2. This pathway has estimated that the method tends to consider an overestimation through the PP2aIA, due to the non-specific inhibition by components present in the matrices of different endemic species (bivalves), that even at low concentrations (<LOD) produce an effect similar to the inhibition produced by OA. This could be related to the presence of some soluble lipids in methanol, which exert a non-specific inhibitory effect on PP2a or the interaction between the matrix, reagents and/or the temperature step involved in the assay [97, 124, 129].

Furthermore, the high concentrations of matrix from endemic species (shellfish) interfere with the PP2aIA. This is due to the colouring that most of the different shellfish matrices have an effect that, without use of appropriate controls, could erroneously establish the presence of toxins associated with the OA-group [97, 111].

Another limiting factor with this assay is its storage temperature. This is because the substrate is very stable at temperatures <15°C and very sensitive to high temperatures (≥37°C), which may lead to underestimation of toxins related to the OA-group. Within this process, it is possible to consider the percentage of recoverability from the different matrices, which has been established between ≈70 and ≈163%, notably affecting the quantification of toxins from shellfish [123].

Finally, one of the biggest disadvantages of PP2aIA is its inability to identify toxins or toxic analogues associated with PTX-, AZA- and YTX-groups [121]. Thus, PP2aIA is only limited to identify the presence of the toxins associated with OA-group, when applicable, even though the co-occurrence of toxins in water or shellfish is evident [12, 13, 26]. Therefore, to identify all toxins in the hypothetical simultaneous presence of all toxic groups in contaminated shellfish, the specific use of enzymatic or cellular assays will be required, considering also that they will never be able to identify the analogues involved in the toxic event accurately [7, 10].

In conclusion, this assay does not tend to be the most adequate to consider the simultaneity of processes of toxic blooms at a global level, in addition to the fact it is not able to provide any information on the profile of the toxins involved in the bloom or contamination of the endemic species (shellfish), which is a crucial stage if we consider that it is the basis of the information in the establishment of risk assessment and management [7].

2.1.3 Chemical method: liquid chromatography coupled with tandem mass spectrometry (LC-MS/MS)

The emergence of instrumental analytical methods for the detection of LMTs responds to the different questions regarding the MBA and PP2aIA. Liquid chromatography coupled with tandem mass spectrometry (LC-MS/MS) is one of the methods that has been highlighted in the last 15 years [130]. This method allows for the quantification of all analogues associated with LMTs (OA-, PTX-, AZA- and YTX-group) when compared to a certified standard for toxins (**Figure 8**) [7, 104].

This chemical method (LC-MS/MS) has been validated and approved by the European Regulation as the new official method for the control of LMTs in shellfish [29, 78, 107, 131]. However, the quantification of toxin analogues is not sufficient for the control and regulatory decision making, since the different analogues have different toxic potencies. For this evaluation, it is necessary to know the relative toxicities of the components making up the mixture of toxins present in a contaminated matrix. For this reason, the term Toxicity Equivalency Factors (TEFs) has been applied, which is defined as the toxicity ratio of a compound from a chemical group that shares the same mode of action of a reference compound in the same group [14, 94, 95, 104].

Thus, the establishment of TEFs in alternative methods to the MBA and PP2aIA for the detection of marine toxins allows, to a great extent, better protection of the consumer in the surveillance programs, since the toxic potential in a mixture of toxins in different endemic species and biological matrices can be better estimated [94, 97, 102, 104].

In this way, the toxin content detected in the different endemic species matrices is expressed as the sum of the equivalents established for each group, considering that for the case of the OA-group, it is necessary to estimate the concentrations of esterified toxins present in the matrices (DTX-3 and acyl-derivatives toxins), a process that requires the evaluation of post-hydrolysis samples of the extract from each matrix [14].

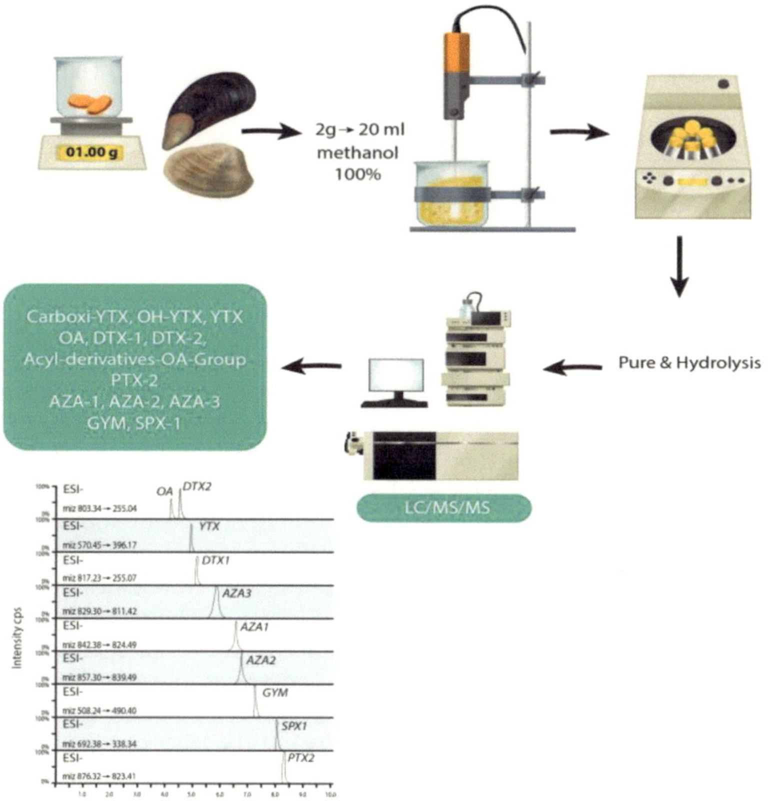

Figure 8.
Liquid chromatography coupled with tandem mass spectrometry (LC-MS/MS) for the detection of lipophilic marine toxins [104, 107, 131].

LC-MS/MS is a method 10 times more sensitive (LOD ≈ 3.0 μg kg^{-1}) than the known 9-anthryldiazomethane (ADAM) fluorescence method (LOD of ≈30 μg kg^{-1}), a chemical method that was initially used to determine OA-group toxins [132, 133] and then proposed to determine toxins of the AZA-group [134]. Nevertheless, for both cases, laborious derivatization steps that increase uncertainty in the analysis process are required. In addition, the ADAM reagent is unstable and should be stored at low temperatures (≤4°C), because its decomposition may induce an incomplete derivatization, interfering with the final analysis [107].

For the use of this analytical chemical method (LC-MS/MS), it is recommended that the analysis of LMTs is started with an initial weight of 150 g of shellfish (without shells), which corresponds to approximately 20–30 bivalves, in order to ensure a representative sample is evaluated [14].

The main advantages of the method include the following [7, 10, 14, 40, 96]:

It is highly specific and sensitive.

- It can individually classify and quantify all toxins of LMTs.
- It provides information on the profiles of all LMTs in the samples.
- It can be automated.
- It provides a reproducible interpretation of the results obtained from the analysis.

- It has no ethical restrictions.

The main disadvantages of the LC-MS/MS method include the following:

- It requires costly equipment and highly trained personnel.
- It requires a wide range of reference standards for identification and quantification.
- The different biological matrices (endemic species) can produce problems in the interpretation of data.

It should be noted that as of July 2011, Regulation (EU) NO. 15/2011 amended the Regulation (EC) No. 2074/2005 in relation to the recognised testing methods for the detection of LMTs in live bivalve molluscs, which establishes the LC-MS/MS method as the official reference method for the detection of LMTs and their use as a routine matter, both for the purposes of official controls at any stage of the food chain, and for the self-controls established by food business operators [29, 30, 78].

All the above mentioned considerations have been established by most countries to improve their quality controls in water and in bivalves from culture centres or endemic species, including water analysis in the location where these cultures are carried out. It should be taken into consideration that the dinoflagellates producers of OA- and/or PTX-, AZA- and YTX-groups at concentrations of ≈2000 cells L^{-1} allow for an adequate accumulation of toxins in bivalves, allowing for the fact that they can be toxic. However, the LOD (sensitivity) of the bioassays for the OA-group in molluscs corresponds to ≈200 μg kg^{-1} shellfish, a value above the internationally established standard of 160 μg OA equivalents kg^{-1} shellfish meat [40].

For the evaluation of all LMTs, the chemical method requires each group to be evaluated in different ionisation modes (**Table 3**) [12, 130, 131, 135].

These ionisation modes allow for the obtainment of suitable ionic fragments and characteristics of each toxic analogue, which allows for the optimal identification of them (**Figure 9**). Nevertheless, it is possible to use both ionisation modes (positive and/or negative mode) for the detection of toxins, but this affects three important

Compound	Ion polarity mode	Precursor ion (m/z)	Product ion 1	Product ion 2	TEF
OA	Negative	803.45	785.4	255.1	1.0
DTX-1	Negative	817.5	785.4	255.1	1.0
DTX-2	Negative	803.5	785.4	255.1	0.6
7-O-palmitoyl-OA	Negative	1041.6	785.4	255.1	1.0*
7-O-palmitoyl-DTX-1	Negative	1055.7	785.4	255.1	1.0*
7-O-palmitoyl-DTX-2	Negative	1041.6	785.4	255.1	1.0*
PTX-1	Positive	892.5	821.5	213.2	1.0
PTX-2	Positive	876.5	823.4	213.2	1.0
PTX-2SA	Positive	894.5	823.4	213.2	1.0
7-epi-PTX2SA	Positive	894.5	823.4	213.3	1.0
PTX-11	Positive	892.5	551.2	213.3	1.0
AZA-1	Positive	842.5	824.5	806.5	1.0
AZA-2	Positive	856.5	838.5	820.5	1.8
AZA-3	Positive	828.5	810.5	792.5	1.4
YTX	Negative	1141.5	1061.7	855.5	1.0
45-OH-YTX	Negative	1157.5	1077.7	871.5	1.0
Homo-YTX	Negative	1155.5	1075.5	869.5	1.0
45-OH-Homo-YTX	Negative	1071.5	1091.5	869.5	0.5

*The TEF of the hydrolysis product of OA, DTX1 or DTX2 would apply [104].

Table 3.
*LC-MS/MS Parameters for determination of lipophilic marine toxins [14, 40, 46, 113, 131, 136, 137]. *The TEF of the hydrolysis product of OA, DTX1 or DTX2 would apply [104].*

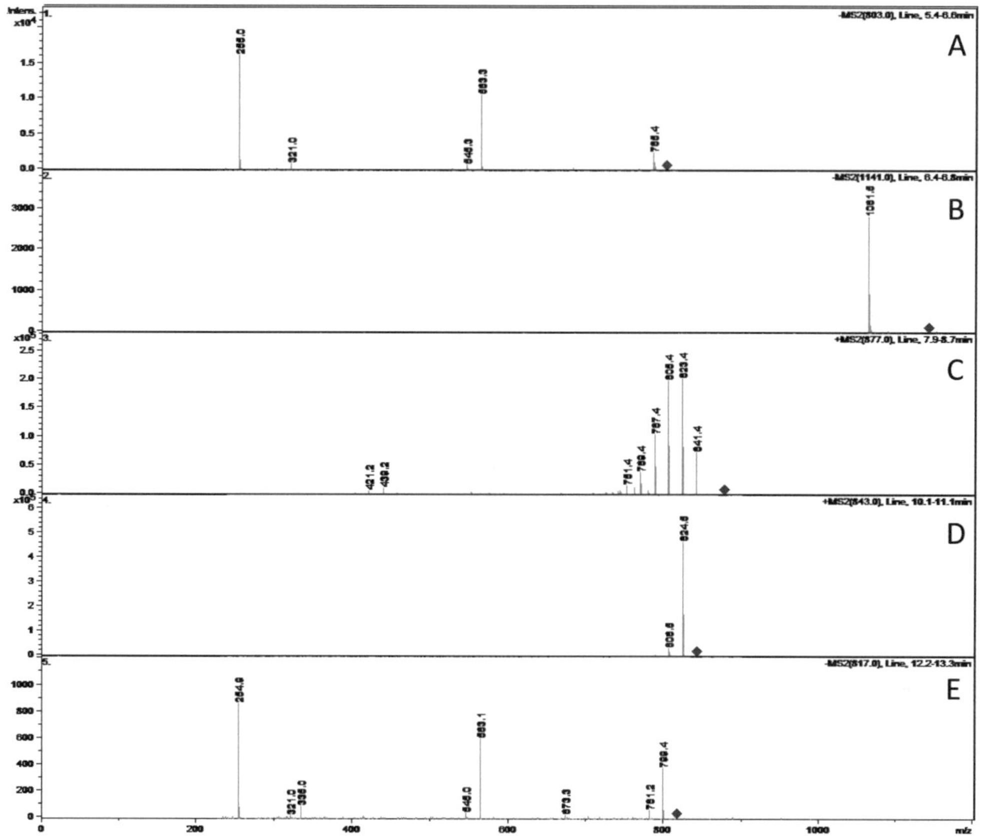

Figure 9.
Mass/mass spectra obtained from a mixture of analytical standards for determination in negative ion mode: AO (A), YTX (B) and DTX-1 (E), and in positive ion mode: PTX-2 (C) and AZA-1 (D) [40, 42, 50].

factors: sensitivity, intensity in the detection of analogues and formation of structural fragments generated by the original toxin, producing a difficult interpretation of the toxic analogues present in the matrices [97, 138]. Thus, the determination of the YTX-group in positive mode produces a very low sensitivity of the ions obtained ($[M{-}2Na + 3H]^+$) making its practical use unfeasible. While in the case of OA-group, the evaluation of post-hydrolysis matrices in positive mode tends to produce matrix effects that do not allow adequate quantification of toxins related to acyl derivatives [139]. In addition, when the base analogues of the OA-group (OA, DTX-1 and DTX-2) are evaluated, the primary ions, through a positive ionisation, tend to always produce a greater variability, resulting in an overestimation of toxicity under this ionisation mode [7].

Another important factor is the mobile phase to be used. Mobile phases modified with different weak carboxylic acids (formic acid, acetic acid, propionic acid and n-butyric acid) may, in some cases, (formic acid) result in decreased electrospray ionisation (ESI) responses of the negative ions produced from toxins. This is consistent with the idea that acidic conditions decrease the negative-ion ESI response, i.e., the conditions at a low pH do not favour the formation of deprotonated analytes [140].

Given the high variability according to the evaluation from different marine matrices in different endemic species of the world, it is necessary to consider the following validation parameters: accuracy, trueness, precision, linearity, robustness, calibration curve check, limit of detection (LOD), limit of quantification (LOQ), and blank quality control [141–143].

3. Discussion

The use of any of the established methods should always consider the variable clearance rates associated with each potentially toxic marine species; those parameters are species-specific in endemic species, since neither body size nor age play a decisive role in the clearance rate of toxins [96]. Hence, it is extremely important to determine all the groups associated with LMTs.

Moreover, the variability of the results obtained with any of the indicated methods is always conditioned to the following parameters:

Growth conditions: The origin of shellfish from endemic species, on the seabed or from cultures, subtidal or intertidal growth, water depth, and water column mix [7].

Clearance rate: Feeding status, species-specific filtration rates and selectivity, as well as the species of microorganisms that can affect endemic species (bacteria and pathogenic viruses) [12, 26].

Metabolism/detoxification: Species-specific differences in endemic species, metabolic changes in bivalves due to seasonal variation, reproductive status and environmental stress [13].

Thus, in order to ensure seafood safety and minimise potential risks to human health, the development of rapid, sensitive and reliable methods to detect different groups in a preventive manner has been proposed. However, two important factors must be considered:

1. Biological and biochemical methods only allow for the detection of specific groups of toxins, and may exclude some, which may cause a potential toxicity when consuming shellfish seemingly free of toxins [14, 102].

2. EFSA has proposed the reduction of maximum limits of all LMTs associated groups, from 160 μg kg $^{-1}$ to 45 μg OA equivalents kg^{-1} (OA-Group); 30 μg AZA equivalents kg^{-1} (AZA-group) and 120 μg PTX equivalents kg^{-1} (PTX-group); these new limits could only be detected completely by using the LC-MS/MS method [7, 14, 107].

4. Conclusions

The above shows that the MBA and PP2aIA assays for LMTs detection cannot be effectively used to follow the toxic variability in molluscs or endemic species in a quantitative manner, including their detoxification stage, as these assays do not allow producers (shellfish growers) to evaluate the results obtained in order to plan their production activities adequately and in advance, since the use of these assays as screening would have less specificity and a higher cost if compared to the confirmation method (LC-MS/MS) used by marine product importing countries.

The LC-MS/MS method does not show any ambiguous results, since it solves and determines the toxic profiles in different toxic marine endemic species (matrices). In addition, it allows for the exact quantification of each group of toxins by keeping those samples under the established legal limits in the market, which by using other methods could prove to be positive (MBA) or not to be able to establish the toxic group involved in a HABs or toxic processes in shellfish (PP2aIA). Thus, the great advantage of LC-MS/MS is the possibility of differentiating toxins belonging to the OA-, YTX-, AZA and PTX- groups, allowing for the evaluation of the potential health risks through the consumption of shellfish.

Acknowledgements

This study was funded by CONICYT/FONDECYT-REGULAR No. 1160168 (granted to C. García).

Conflict of interest

The authors declare no conflict of interest.

Author details

Carlos García[1,2]*, Javiera Oyaneder-Terrazas[1] and Héctor R. Contreras[3]

1 Laboratory of Marine Toxins, Physiology and Biophysics Program, Faculty of Medicine, University of Chile, Santiago, Chile

2 Environment Department, Foundation for Human and Environmental Sciences Research, Santiago, Chile

3 Department of Basic and Clinical Oncology, Faculty of Medicine, University of Chile, Santiago, Chile

*Address all correspondence to: cgarcia@med.uchile.cl

References

[1] James KJ, Carey B, O'Halloran J, Van Pelt FNAM, Skrabakova Z. Shellfish toxicity: Human health implications of marine algal toxins. Epidemiology and Infection. 2010;**138**:927-940

[2] Fux E, Smith JL, Tong M, Guzman L, Anderson DM. Toxin profiles of five geographical isolates of *Dinophysis* spp. from North and South America. Toxicon. 2011;**57**:275-287

[3] Díaz PA, Reguera B, Ruiz-Villarreal M, Pazos Y, Velo-Suárez L, Berger H, et al. Climate variability and oceanographic settings associated with interannual variability in the initiation of *Dinophysis acuminata* blooms. Marine Drugs. 2013;**11**:2964-2981

[4] Reguera B, Riobó P, Rodríguez F, Díaz PA, Pizarro G, Paz B, et al. Dinophysis toxins: Causative organisms, distribution and fate in shellfish. Marine Drugs. 2014;**12**:394-461

[5] Prego-Faraldo MV, Valdiglesias V, Laffon V, Mendez J, Eirin-Lopez JM. Early genotoxic and cytotoxic effects of the toxic dinoflagellate *Prorocentrum lima* in the mussel *Mytilus galloprovincialis*. Toxins. 2016;**8**:159

[6] Hallegraeff GM. Ocean climate change, phytoplankton community responses, and harmful algal blooms: A formidable predictive challenge. Journal of Phycology. 2010;**46**:220-235

[7] Hess P. Requirements for screening and confirmatory methods for the detection and quantification of marine biotoxins in end product and official control. Analytical and Bioanalytical Chemistry. 2010;**397**:1683-1694

[8] FAO/IOC/WHO. Report of the Joint FAO/IOC/WHO Ad Hoc Expert Consultation on Biotoxins in Bivalve Molluscs. 2004. Available from: ftp://ftp.fao.org/es/esn/food/biotoxin_-report_en.pdf [Accessed: Apr 4, 2012]

[9] Etheridge SM, Roesler CS. Effects of temperature, irradiance, and salinity on photosynthesis, growth rates, total toxicity, and toxin composition for *Alexandrium fundyense* isolates from the Gulf of Maine and Bay of Fundy. Deep-Sea Research II. 2005;**52**:2491-2500

[10] Twiner MJ, Rehmann N, Hess P, Doucette GJ. Azaspiracid shellfish poisoning: A review on the chemistry, ecology and toxicology with an emphasis on human health impact. Marine Drugs. 2008;**6**:39-72

[11] Navarro JM, Muñoz MG, Contreras AM. Temperature as a factor regulating growth and toxin content in the dinoflagellate *Alexandrium catenella*. Harmful Algae. 2006;**5**:762-769

[12] Zamorano R, Marín M, Cabrera F, Figueroa D, Contreras C, Barriga A, et al. Determination of the variability of both hydrophilic and lipophilic toxins in endemic wild bivalves and carnivorous gastropods from the Southern part of Chile. Food Additives & Contaminants: Part A. 2013;**30**:1660-1677

[13] García C, Pérez F, Contreras C, Figueroa D, Barriga A, López-Rivera A, et al. Saxitoxins and okadaic acid group: Accumulation and distribution in invertebrate marine vectors from Southern Chile. Food Additives & Contaminants: Part A. 2015;**32**:984-1002

[14] EFSA. Opinion of the scientific panel on contaminants in the food chain on a request from the European commission on marine biotoxins in shellfish—Okadaic acid and analogues. EFSA Journal. 2008;**589**:1-62

[15] Shumway S. Phycotoxin-related shellfish poisoning: Bivalve molluscs are not the only vectors. Reviews in Fisheries Science. 1995;**3**:1-31

[16] Jauffrais T, Marcaillou C, Herrenknecht C, Truquet P, Séchet V, Nicolau E, et al. Azaspiracid accumulation, detoxification and biotransformation in blue mussels (*Mytilus edulis*) experimentally fed *Azadinium spinosum*. Toxicon. 2012;**60**:582-595

[17] García C, Seguel M, Uribe JC. Dynamics of toxic dinoflagellates' blooms in the Austral Pacific Region: Distribution, toxicity and impact on aquaculture. In: Tobias RD, Vermette M, editors. Dinoflagellates: Biology, Geographical Distribution and Economic Importance. New York, NY, USA: Nova Science Publishers; 2013. pp. 1-56

[18] Rossini GP, Hess P. Phycotoxins: Chemistry, mechanisms of action and shellfish poisoning. Experientia Supplementum (EXS) Journal. 2010;**100**:65-122

[19] García C, Rodriguez-Unda N, Contreras C, Barriga A, Lagos N. Lipophilic toxin profiles detected in farmed and benthic mussels populations from the most relevant production zones in Southern Chile. Food Additives & Contaminants: Part A. 2012;**29**:1011-1020

[20] Deeds J, Landsberg J, Etheridge S, Pitcher G, Longan S. Non-traditional vectors for paralytic shellfish poisoning. Marine Drugs. 2008;**6**:308-348

[21] Lopes VM, Lopes AR, Costa P, Rosa R. Cephalopods as vectors of harmful algal bloom toxins in marine food webs. Marine Drugs. 2013;**11**:3381-3409

[22] May SP, Burkholder JAM, Shumway SE, Wikfors GH, Frank Dasiaaret H, Dorch Q. Effects of the toxic dinoflagellate *Alexandrium monilatum* on survival, grazing and behavioral response of three ecologically important shellfish species. Harmful Algae. 2010;**9**:281-293

[23] McCarron P, Hess P. Tissue distribution and effects of heat treatments on the content of domoic acid in blue mussels, *Mytilus edulis*. Toxicon. 2006;**47**:473-479

[24] Etheridge SM. Paralytic shellfish poisoning: Seafood safety and human health perspectives. Toxicon. 2010;**56**:108-122

[25] Morgan KL, Larkin SL, Adams CM. Firm level economic effects of HABs: A tool for business loss assessment. Harmful Algae. 2009;**8**:212-218

[26] García C, Oyaneder-Terrazas J, Contreras C, del Campo M, Torres R, Contreras HR. Determination of the toxic variability of lipophilic biotoxins in marine bivalve and gastropod tissues treated with an industrial canning process. Food Additives & Contaminants: Part A. 2016;**33**:1711-1727

[27] Glibert PM, Burkholder JM, Kana TM. Recent insights about relationships between nutrient availability, stoichiometry, and the distribution and food web effects of pelagic and benthic Prorocentrum species. Harmful Algae. 2012;**14**:231-259

[28] Dominguez HJ, Paz B, Daranas AH, Norte M, Franco JM, Fernández JJ. Dinoflagellate polyether within the yessotoxin, pectenotoxin and okadaic acid toxin groups: characterization, analysis and human health implications. Toxicon. 2010;**56**:191-217

[29] EU, European Commission, COMMISSION REGULATION (EU) No. 15/2011 of 10 January 2011 amending regulation (EC) No. 2074/2005 as regards recognized testing methods for detecting marine biotoxins in live bivalve molluscs. Official Journal European Union. 2011. L 6/3 -L6/6. 11.01.11

[30] Paredes I, Rietjens IM, Vieites JM, Cabado AG. Update of risk assessments of main marine biotoxins

in the European Union. Toxicon. 2011;**58**:336-354

[31] García C, Contreras HR. Effects of both paralytic shellfish toxins and diarrhetic shellfish toxins in human poisoning: toxicity, distribution and biotransformation. In: Hay RM, editor. Shellfish Human Consumption, Health Implications and Conservation Concerns. New York, NY, USA: Nova Science Publishers; 2014. pp. 345-384

[32] Aune T, Espenes A, Bunæs Aasen JA, Quilliam MA, Hess P, Larsen S. Study of possible combined toxic effects of azaspiracid-1 and okadaic acid in mice via the oral route. Toxicon. 2012;**60**:895-906

[33] Yasumoto T, Murata M, Oshima Y, Sano M, Matsumoto G, Clardy J. Diarrhetic shellfish toxins. Tetrahedron. 1985;**41**:1019-1025

[34] Vieira AC, Rubiolo JA, López-Alonso H, Cifuentes JM, Alfonso A, Bermúdez R, et al. Oral toxicity of okadaic acid in mice: Study of lethality, organ damage, distribution and effects on detoxifying gene expression. Toxins. 2013;**5**:2093-2108

[35] Paz B, Daranas AH, Norte M, Riobó P, Franco JM, Fernández JJ. Yessotoxins, a group of marine polyether toxins: An overview. Marine Drugs. 2008;**6**:73-102

[36] Gerssen A, Pol-Hofstad I, Poelman M, Mulder P, van, den Top H, de Boer J. Marine toxins: Chemistry, toxicity, occurrence and detection, with special reference to the Dutch situation. Toxicon. 2010;**2**:878-904

[37] Manfrin C, Dreos R, Battistella S, Beran A, Gerdol M, Varotto L, et al. Mediterranean mussel gene expression profile induced by okadaic acid exposure. Environmental Science & Technology. 2010;**44**:8276-8283

[38] Lindahl O, Lundve B, Johansen M. Toxicity of *Dinophysis* spp. in relation to population density and environmental conditions on the Swedish west coast. Harmful Algae. 2007;**6**:218-231

[39] García C, Gonzalez V, Cornejo C, Palma-Fleming H, Lagos N. First evidence of dinophisistoxin-1 and carcinogenic polycyclic aromatic hydrocarbons in smoked bivalves collected in the Patagonia fjords. Toxicon. 2004;**43**:121-131

[40] García, C. Detection and quantification of lipophilic marine biotoxins by liquid chromatography tandem mass spectrometry (LC-MS/MS) from endemic species (Mytilus sp.) and Gastropods from Southern Chile. In: Gray D, editor. Marine Toxins: Detection Methods, Chemical and Biological Aspects and Health Effects. New York: Nova Science Publishers; 2016. p. 1-32

[41] Suzuki T, Mackenzie L, Stirling D, Adamson J. Pectenotoxin-2 seco acid: A toxin converted from pectenotoxin-2 by the New Zealand greenshell mussel *Perna canaliculus*. Toxicon. 2001;**39**:507-514

[42] García C, Pruzzo M, Rodriguez-Unda N, Contreras C, Lagos N. First evidence of okadaic acid acyl-derivative and dinophysistoxin-3 in mussel samples collected in Chiloe island, southern Chile. The Journal of Toxicological Sciences. 2010;**35**:335-344

[43] Vale PJ, Botelho MJ, Rodrigues SM, Gomes SS, Sampayo MA. Two decades of marine biotoxin monitoring in bivalves from Portugal (1986-2006): A review of exposure assessment. Harmful Algae. 2008;**7**:11-25

[44] Torgersen T, Miles C, Rundberget T, Wilkins A. New esters of okadaic acid in seawater and blue mussels (*Mytilus edulis*). Journal of Agricultural and Food Chemistry. 2008;**56**:9628-9635

[45] Torgersen T, Sandvik M, Lundve B, Lindegarth S. Profiles and levels of fatty acid esters of okadaic acid Group Toxins and pectenotoxins during toxin depuration. Part II: Blue mussels (*Mytilus edulis*) and flat oyster (*Ostrea edulis*). Toxicon. 2008;**52**:418-427

[46] Suzuki T, Quilliam M. LC-MS/MS analysis of diarrhetic shellfish poisoning (DSP) toxins, okadaic acid and dinophysistoxin analogues, and other lipophilic toxins. Analytical Sciences. 2011;**27**:571-584

[47] Prassopoulou E, Katikou P, Georgantelis D, Kyritsakis A. Detection of okadaic acid and related esters in mussels during diarrhetic shellfish poisoning (DSP) episodes in Greece using the mouse bioassay, the PP2A inhibition assay and HPLC with fluorimetric detection. Toxicon. 2009;**53**:214-227

[48] Svensson S. Depuration of okadaic acid (diarrhetic shellfish toxins) in mussels, *Mytilus edulis* (Linnaeus), feeding on different quantities of nontoxic algae. Aquaculture. 2003;**218**:277-291

[49] García C, Truan D, Lagos M, Santelices JP, Díaz JC, Lagos N. Metabolic transformation of dinophysistoxin-3 into dinophysistoxin-1 causes human intoxication by consumption of O-acyl-derivatives dinophysistoxins contaminated shellfish. The Journal of Toxicological Sciences. 2005;**30**:287-296

[50] García C, Schonstedt V, Santelices J, Lagos N. Hight amount of Diniphysistoxin-3 in *Mytilus chilensis* collected in seno de Reloncavi, Chile, during massive human intoxication associated with outbreak of *Vibrio parahaemolyticus*. The Journal of Toxicological Sciences. 2006;**31**:305-314

[51] Hamano Y, Kinoshita Y, Yasumoto T. Enteropathogenicity of diarrhetic shellfish toxins in intestinal models. Journal of the Food Hygienic Society of Japan. 1986;**27**:375-379

[52] Fujiki H, Sueoka E, Suganuma M. Tumor promoters: From chemicals to inflammatory proteins. Journal of Cancer Research and Clinical Oncology. 2013;**139**:1603-1614

[53] López-Rodas V, Maneiro E, Martínez J, Navarro M, Costa E. Harmful algal blooms, red tides and human health: Diarrhetic shellfish poisoning and colorectal cancer. Anales de la Real Academia Nacional de Farmacia. 2006;**72**:391-408

[54] Maneiro E, Victorio R, Costas E, Hernandez JM. Shellfish consumption: A major risk factor for colorectal cancer. Medical Hypotheses. 2008;**70**:409-412

[55] Li Z, Mengmeng G, Shouguo Y, Qingyin W, Zhijun T. Investigation of pectenotoxins profiles in the Yellow Sea (China) using a passive sampling technique. Marine Drugs. 2010;**8**:1263-1272

[56] Miles CO. Pectenotoxins. In: Botana L, editor. Phycotoxins: Chemistry and Biochemistry. Oxford, UK: Blackwell Publishing; 2007. pp. 159-186

[57] EFSA. Scientific opinion of the panel on contaminants in the food chain on a request from the European commission on marine biotoxins in shellfish—Pectenotoxin group. EFSA Journal. 2009;**1109**:1-47

[58] Krock B, Seguel CG, Valderrama K, Tillmann U. Pectenotoxins and yessotoxin from Arica Bay, north Chile as determined by tandem mass spectrometry. Toxicon. 2009;**54**:364-367

[59] Fabro E, Almandoz GO, Ferrario ME, Hoffmeyer MS, Pettigrosso RE, Uibrig R, et al. Co-occurrence of *Dinophysis tripos* and pectenotoxins in Argentinean shelf waters. Harmful Algae. 2015;**42**:25-33

[60] Miles CO, Wilkins AL, Munday R, Dines MH, Hawkes AD, Briggs LR, et al. Isolation of pectenotoxin-2 from *Dinophysis acuta* and its conversion to pectenotoxin-2 seco acid, and preliminary assessment of their acute toxicities. Toxicon. 2004;**43**:1-9

[61] Miles CO. Identification of fatty acid esters of pectenotoxin-2 seco acid in Blue mussels (*Mytilus edulis*) from Ireland. Journal of Agricultural and Food Chemistry. 2006;**54**:5672-5678

[62] Aasen J, Hardstaff WR, Aune T, Quilliam MA. Discovery of fatty acid ester metabolites of spirolide toxins in mussels from Norway using liquid chromatography/tandem mass spectrometry. Rapid Communications in Mass Spectrometry. 2006;**20**:1531-1537

[63] Jung IH, Sim CJ, Lee CO. Cytotoxic compounds from a two sponge association. Journal of Natural Products. 1995;**58**:1722-1726

[64] Satake M, MacKenzie AL, Yasumoto T. Identification of *Protoceratium reticulatum* as the biogenic origin of yessotoxin. Natural Toxins. 1997;**5**:164-167

[65] Tubaro A, Sidari L, Della Loggia R, Yasumoto T. Occurrence of homoyessotoxin in phytoplankton and mussels from Northern Adriatic Sea. In: Reguera B, Blanco J, Fernandez ML, Wyatt T, editors. Harmful Algae, Xunta de Galicia and Intergovernmental Oceanographic Commission of UNESCO. Santiago de Compostela: Grafisant; 1998. pp. 470-472

[66] Rhodes L, McNabb P, de Salas M, Briggs L, Beuzenberg V, Gladstone M. Yessotoxin production by *Gonyaulax spinifera*. Harmful Algae. 2006;**5**:148-155

[67] Sosa S, Ardizzone M, Beltramo D, Vita F, Dell'Ovo V, Barreras A, et al. Repeated oral co-exposure to yessotoxin and okadaic acid: A short term toxicity study in mice. Toxicon. 2013;**76**:94-102

[68] Álvarez G, Uribe E, Díaz R, Braun M, Mariño C, Blanco J. Bloom of the yessotoxin producing dinoflagellate *Protoceratium reticulatum* (Dinoflagellata) in northern Chile. Journal of Sea Research. 2011;**65**:427-434

[69] Paz B, Blanco J, Franco JM. Yessotoxins production during the culture of *Protoceratium reticulatum* strains isolated from Galician Rias Baixas (NW Spain). Harmful Algae. 2013;**21-22**:13-19

[70] Akselman R, Krock B, Alpermann TJ, Tillmann U, Borel CM, Almandoz GO, et al. *Protoceratium reticulatum* (Dinophyceae) in the austral Southwestern Atlantic and the first report on YTX-production in shelf waters of Argentina. Harmful Algae. 2015;**45**:40-52

[71] Tubaro A, Sosa S, Carbonatto M, Altinier G, Vita F, Melato M, et al. Oral and intraperitoneal acute toxicity studies of yessotoxin and homoyessotoxins in mice. Toxicon. 2003;**41**:783-792

[72] Tubaro A, Sosa S, Altinier G, Soranzo MR, Satake M, Loggia RD, et al. Short-term oral toxicity of homoyessotoxins, yessotoxin and okadaic acid in mice. Toxicon. 2004;**43**:439-445

[73] Visciano P, Schirone M, Tofalo R, Berti M, Luciani M, Ferri N, et al. Detection of yessotoxin by three different methods in *Mytilus galloprovincialis* of Adriatic Sea, Italy. Chemosphere. 2013;**90**:1077-1082

[74] Ciminiello P, Dell-Aversano C, Fattorusso E, Forino M, Grauso L, Magno S, et al. Desulfoyessotoxins from Adriatic mussels: A new problem for seafood safety control. Chemical Research in Toxicology. 2007;**20**:95-98

[75] Tubaro A, Gangaspero A, Ardizzone M, Soranzo MR, Vita F, Yasumoto T, et al. Ultrastructural damage to heart tissue from repeated oral exposure to yessotoxin resolves in 3 months. Toxicon. 2008;**51**:1225-1235

[76] Aune T, Sorby R, Yasumoto T, Ramstad H, Landsverk T. Comparison of oral and intraperitoneal toxicity of yessotoxin towards mice. Toxicon. 2002;**40**:77-82

[77] Aune T, Aasen JA, Miles CO, Larsen S. Effect of mouse strain and gender on LD_{50} of yessotoxin. Toxicon. 2008;**52**:535-540

[78] European Commission. COMMISSION REGULATION (EU) No. 786/2013 of 16 August 2013 amending annex III to regulation (EC) No. 853/2004 of the European Parliament and of the Council as regards the permitted limits of yessotoxins in live bivalve molluscs. Official Journal European Union. 2013. L 220/14

[79] Tillman U, Elbrachter M, Krock B, John U, Cembella A. *Azadinium spinosum* gen. et sp. nov. (Dinophyceae) identified as primary producer of azaspiracid toxins. European Journal of Phycology. 2009;**44**:63-79

[80] Potvin E, Jeong H, Seon Kang N, Tillmann U, Krock B. First report of the photosynthetic dinoflagellate genus azadinium in the Pacific Ocean: Morphology and molecular characterization of *Azadinium cf. poporum*. The Journal of Eukaryotic Microbiology. 2012;**59**:145-156

[81] O'Driscoll D, Skrabakova Z, James KJ. Confirmation of extensive natural distribution of azaspiracids in the tissue compartments of mussels (*Mytilus edulis*). Toxicon. 2014;**92**:123-128

[82] Krock B, Tillmann U, John U, Cembella AD. Characterization of azaspiracids in plankton size fractions and isolation of an azaspiracid producing dinoflagellate from the North Sea. Harmful Algae. 2009;**8**:254-263

[83] Amzil Z, Sibat M, Royer F, Savar V. First report of azaspiracid and yessotoxin groups detection in French shellfish. Toxicon. 2008;**52**:39-48

[84] Rehmann N, Hess P, Quilliam MA. Discovery of new analogs of the marine biotoxin azaspiracid in blue mussels *Mytilus edulis* by ultraperformance liquid chromatography/tandem mass spectrometry. Rapid Communications in Mass Spectrometry. 2008;**22**:549-558

[85] Picot C, Nguyen TA, Roudot AC, Parent-Massin D. A preliminary risk assessment of human exposure to phycotoxins in shellfish: A review. Human and Ecological Risk Assessment: An International Journal. 2011;**17**:328-366

[86] Twiner MJ, Hess P, Dechraoui MYB, McMahon T, Samons MS, Satake M, et al. Cytotoxic and cytoskeletal effects of azaspiracid-1 on mammalian cell lines. Toxicon. 2005;**45**:891-900

[87] Ronzitti G, Hess P, Rehmann N, Rossini GP. Azaspiracid-1 alters the E-cadherin pool in epithelial cells. Toxicological Sciences. 2007;**95**:427-435

[88] James K, O'Driscoll D, Fernandez M, Furey A. Azaspiracids: Chemistry, bioconversion and determination. In: Botana LM, editor. Seafood and Freshwater Toxins: Pharmacology, Physiology and Detection. New York: CRC Press, Taylor and Francis; 2008. pp. 763-773

[89] Ito E, Satake M. Azaspiracid, a new marine toxin isolated from mussels: chemistry and histopathology. In: Fingerman M, Nagabhushanam R, editors. Recent Advances in Marine Biotechnology. Seafood Safety and Human Health. Enfield, New Hampshire: Science Publishers; 2002. pp. 31-39

[90] Ito E, Satake M, Ofuji K, Higashi M, Harigaya K, McMahon T, et al. Chronic effects in mice caused by oral administration of sublethal doses of azaspiracid, a new marine toxin isolated from mussels. Toxicon. 2002;**40**:193-203

[91] EFSA. Scientific opinion of the panel on contaminants in the food chain on a request from the European commission on marine biotoxins in shellfish—Summary on regulated marine biotoxins. EFSA Journal. 2009;**1306**:1-23

[92] Magdalena AB, Lehane M, Moroney C, Furey A, James KJ. Food safety implications of the distribution of azaspiracids in the tissue compartments of scallops (*Pecten maximus*). Food Additives & Contaminants: Part A. 2003;**20**:154-160

[93] Furey A, O'Doherty S, O'Callaghan K, Lehane M, James KJ. Azaspiracids poisioning (AZP) toxins in shellfish: Toxicological and health considerations. Toxicon. 2010;**56**:173-190

[94] Botana LM, Vilariño N, Elliott CT, Campbell K, Alfonso A, Vale C. The problem of toxicity equivalent factors in developing alternative methods to animal bioassays for marine toxin detection. Trends in Analytical Chemistry. 2010;**29**:1316-1325

[95] Campbell K, Vilariño N, Botana LM, Elliott CT. A European perspective on progress in moving away from the mouse bioassay for marine toxin analysis. Trends in Analytical Chemistry. 2011;**30**:2

[96] Prego-Faraldo MV, Valdiglesias V, Méndez J, Eirín-López JM. Okadaic acid meet and greet: An insight into detection methods, response strategies and genotoxic effects in marine invertebrates. Marine Drugs. 2013;**11**:2829-2845

[97] Garibo D, Damaso E, Eixarch H, de la Iglesia P, Fernandez-Tejedor M, Diogene J, et al. Protein phosphatase inhibition assays for okadaic acid detection in shellfish: Matrix effects, applicability and comparison with LC-MS/MS analysis. Harmful Algae. 2012;**19**:68-75

[98] Suzuki H. Influence of body weight of mice on the susceptibility to okadaic acid, a diarrhetic shellfish poisoning toxin. Food Additives & Contaminants: Part A. 2014;**31**:719-722

[99] Matias WG, Traore A, Creppy EE. Variations in the distribution of okadaic acid in organs and biological fluids of mice related to diarrhoeic syndrome. Human & Experimental Toxicology. 1999;**18**:345-350

[100] Yasumoto T, Oshima Y, Yamaguchi M. Occurrence of a new type of shellfish poisoning in the Tohoku district. Bulletin of the Japanese Society of Scientific Fisheries. 1978;**46**:1249-1275

[101] Munday R. Is protein phosphatase inhibition responsible for the toxic effects of okadaic acid in animals? Toxins. 2013;**5**:267-285

[102] Aune T, Larsen S, Aasen JAB, Rehmann N, Satake M, Hess P. Relative toxicity of dinophysistoxin-2 (DTX2) compared with okadaic acid, based on acute intraperitoneal toxicity in mice. Toxicon. 2007;**49**:1-7

[103] Hu T, Curtis JM, Walter JA, Wright JLC. Identification of DTX-4, a new water-soluble phosphatase inhibitor from the toxic dinoflagellate *Prorocentrum lima*. Journal of the Chemical Society, Chemical Communications. 1995;**42**:597-599

[104] Botana LM, Hess P, Munday R, Nathalie A, De Grasse SL, Feeley M, et al. Derivation of toxicity equivalency factors for marine biotoxins associated with bivalve molluscs. Trends in Food Science & Technology. 2017;**50**:15-24

[105] Doucet E, Ross N, Quilliam M. Enzymatic hydrolysis of esterified diarrhetic shellfish poisoning toxins and pectenotoxins. Analytical and Bioanalytical Chemistry. 2007;**389**:335-342

[106] Suzuki T, Yoshizawa R, Kawamura T, Yamasaki M. Interference of free fatty acids from the hepatopancreas of mussels with the mouse bioassay for shellfish toxins. Lipids. 1996;**31**:6

[107] Sassolas A, Hayat A, Catanante G, Marty JL. Detection of the marine toxin okadaic acid: Assessing seafood safety. Talanta. 2013;**105**:306-316

[108] Christian B, Luckas B. Determination of marine biotoxins relevant for regulations: From the mouse bioassay to coupled LC-MS methods. Analytical and Bioanalytical Chemistry. 2008;**391**:117-134

[109] Aasen JAB, Espenes A, Hess P, Aune T. Sub-lethal dosing of azaspiracid-1 in female NMRI mice. Toxicon. 2010;**56**:1419-1425

[110] Ikehara T, Shinjo F, Ikehara S, Imamura S, Yasumoto T. Baculovirus expression, purification, and characterization of human protein phosphatase 2A catalytic subunits α and β. Protein Expression and Purification. 2006;**45**:150-156

[111] Hayat A, Barthelmebs L, Marty JL. A simple colorimetric enzymatic assay for okadaic acid detection based on the immobilization of protein phosphatase 2A in sol-gel. Applied Biochemistry and Biotechnology. 2012;**166**:47-56

[112] Huhn J, Jeffrey PD, Larsen K, Rundberget T, Rise F, Cox NR, et al. A structural basis for the reduced toxicity of dinophysistoxin-2. Chemical Research in Toxicology. 2009;**22**:1782-1786

[113] Konoki K, Onoda T, Watanabe R, Cho Y, Kaga S, Suzuki T, et al. *In Vitro* acylation okadaic acid in the presence of various bivalve's extracts. Marine Drugs. 2013;**11**:300-315

[114] Rubiolo JA, López-Alonso H, Alfonso A, Vega FV, Rodríguez-Vieytes M, Botana LM. Bioengineered protein phosphatase 2A. Bioengineered. 2013;**4**:72-77

[115] Bialojan C, Takai A. Inhibitory effect of a marine sponge toxin, okadaic acid, on protein phosphatases. Specificity and kinetics. The Biochemical Journal. 1988;**256**:283-290

[116] Tubaro A, Florio C, Luxich E, Sosa S, Loggia LD, Yasumoto T. A protein phosphatase 2A inhibition assay for a fast and sensitive assessment of okadaic acid contamination in mussels. Toxicon. 1996;**34**:743-752

[117] Zolnierowicz S, Van Hoof C, Andjelkovic N, Cron P, Stevens I, Merlevede W, et al. The variable subunit associated with protein phosphatase 2A0 defines a novel multimember family of regulatory subunits. The Biochemical Journal. 1996;**317**:187-194

[118] Kamibayashi C, Estes R, Lickteig RL, Yang SI, Craft C, Mumby MC. Comparison of heterotrimeric protein phosphatase 2A containing different B subunits. The Journal of Biological Chemistry. 1994;**269**:20139-20148

[119] Wadzinski BE, Eisfelder BJ, Peruski LF, Mumby MC, Johnson GL. NH_2-terminal modification of the phosphatase 2A catalytic subunit allows functional expression in mammalian cells. The Journal of Biological Chemistry. 1992;**267**:16883-16888

[120] Evans DR, Myles T, Hofsteenge J, Hemmings BA. Functional expression of human PP2Ac in yeast permits the identification of novel C-terminal and dominant-negative mutant forms. The Journal of Biological Chemistry. 1999;**274**:24038-24046

[121] Ikehara T, Imamura S, Yoshino A, Yasumoto T. PP2A inhibition assay using recombinant enzyme for rapid detection of okadaic acid and its analogs in shellfish. Toxins. 2010;**2**:195-204

[122] Takai A, Murata M, Torigoe K, Isobe M, Mieskes G, Yasumoto T. Inhibitory effect of okadaic acid derivatives on protein phosphatases. The Biochemical Journal. 1992;**284**:539-544

[123] Smienk H, Calvo D, Razquin P, Domínguez E, Mata L. Single laboratory validation of a ready to use phosphatase inhibition assay for detection of okadaic acid toxins. Toxins. 2012;**4**:339-352

[124] Garibo D, de la Iglesia P, Diogene J, Campas M. Inhibition equivalency factors for dinophysistoxin-1 and dinophysistoxin-2 in protein phosphatase assays: Applicability to the analysis of shellfish samples and comparison with LC-MS/MS. Journal of Agricultural and Food Chemistry. 2013;**61**:2572-2579

[125] Holmes CFB, Luu HA, Carrier F, Schmitz FJ. Inhibition of protein phosphatases-1 and -2A with acanthifolicin. Comparison with diarrhetic shellfish toxins and identification of a region on okadaic acid important for phosphatase inhibition. FEBS Letters. 1990;**270**:216-218

[126] Nishiwaki S, Fujiki H, Suganuma M, Furuya-Suguri H, Matsushima R, Iida Y, et al. Structure-activity relationship within a series of okadaic acid derivatives. Carcinogenesis. 1990;**11**:1837-1841

[127] McNabb P. Chemistry, metabolism, and chemical analysis of okadaic acid group of toxins. In: Botana LM, editor. Seafood and Freshwater Toxins. Pharmacology Physiology and Detection. Boca Raton, FL: CRC Press, Taylor & Francis Group, LLC; 2008. pp. 209-228

[128] Albano C, Ronzitti G, Rossini AM, Callegari F, Rossini GP. The total activity of a mixture of okadaic acid-group compounds can be calculated by those of individual analogues in a phosphoprotein phosphatase 2A assay. Toxicon. 2009;**53**:631-637

[129] Honkanen RE, Mowdy DE, Dickey RW. Detection of DSP-toxins, okadaic acid, and dinophysistoxin-1 in shellfish by serine/threonine protein phosphatase assay. Journal of AOAC International. 1996;**79**:1336-1343

[130] EURLMB. EU-harmonised Standard Operating Procedure for Determination of Lipophilic Marine Biotoxins in Molluscs by LC-MS/MS. Version5. 2015. Available from: http://aesan.msssi.gob.es/CRLMB/docs/docs/metodos_analiticos_de_desarrollo/EU-Harmonised-SOP-LIPO-LCMSMS_Version5.pdf

[131] Gerssen A, Mulder PP, de Boer J. Screening of lipophilic marine toxins in shellfish and algae: Development of a library using liquid chromatography coupled to orbitrap mass spectrometry. Analytica Chimica Acta. 2011;**685**:176-185

[132] Lee JS, Yanagi T, Kenma R, Yasumoto T. Fluorometric determination of diarrhetic shellfish toxins by high performance liquid chromatography. Agricultural and Biological Chemistry. 1987;**51**:877-881

[133] Gerssen A, McElhinney MA, Mulder PPJ, Bire R, Hess P, de Boer J. Solid phase extraction for removal of matrix effects in lipophilic marine toxin analysis by liquid chromatography-tandem mass spectrometry. Analytical and Bioanalytical Chemistry. 2009;**394**:1213-1226

[134] McCarron P, Giddings SD, Miles CO, Quilliam MA. Derivatization of azaspiracid biotoxins for analysis by liquid chromatography

with fluorescence detection. Journal of Chromatography A. 2011;**1218**:8089-8096

[135] Zhang H, Liu W, He X, Liang L, Ding W, He Z. Determination of okadaic acid related toxins from shellfish (*Sinonovacula constricta*) by high performance liquid chromatography tandem mass spectrometry. Agricultural Sciences. 2013;**4**:1-6

[136] Carey B, Fidalgo-Sáez MJ, Hamilton B, O'Halloran J, van Pelt FNAM, James KJ. Elucidation of the mass fragmentation pathways of the polyether marine toxins, dinophysistoxins, and identification of isomer discrimination processes. Rapid Communications in Mass Spectrometry. 2012;**26**:1793-1802

[137] García-Altares M, Diogène J, de, la Iglesia P. The implementation of liquid chromatography tandem mass spectrometry for the official control of lipophilic toxins in seafood: Single-laboratory validation under four chromatographic conditions. Journal of Chromatography A. 2013;**1275**:48-60

[138] Domènech A, Cortés-Francisco N, Palacios O, Franco JM, Riobó P, Llerena JJ, et al. Determination of lipophilic marine toxins in mussels. Quantification and confirmation criteria using high resolution mass spectrometry. Journal of Chromatography A. 2014;**1328**:16-25

[139] Suzuki T, Igarashi T, Ichimi K, Watai M, Suzuki M, Ogiso E, et al. Kinetics of diarrhetic shellfish poisoning toxins, okadaic acid, dinophysistoxin-1, pectenotoxin-6 and yessotoxin in scallops *Patinopecten yessoensis*. Fisheries Science. 2005;**71**:948-955

[140] Wu Z, Gao W, Phelps MA, Wu D, Miller DD, Dalton JT. Favorable effects of weak acids on negative ion electrospray ionization mass spectrometry. Analytical Chemistry. 2004;**76**:839-847

[141] Chen J, Yan T, Xu J, He S, Zhao P, Yan X. Simultaneous determination of toxins in algae and water samples by high performance liquid chromatography with triple quadrupole mass spectrometry. Journal of Separation Science. 2012;**35**:1094-1101

[142] Chapela MJ, Reboreda A, Vieites JM, Cabado AG. Lipophilic toxins analyzed by liquid chromatography mass spectrometry and comparison with mouse bioassay in fresh, frozen, and processed molluscs. Journal of Agricultural and Food Chemistry. 2008;**56**:8979-8986

[143] Smienk H, Dominguez E, Rodriguez-Velasco M, Clarke D, Kapp K, Katikou P, et al. Quantitative determination of the okadaic acid toxins group by a colorimetric phosphatase inhibition assay: Interlaboratory study. Journal of AOAC International. 2013;**96**:77-85

Section 2

Animal Biology

Chapter 7

The Genotyping of Glucose 6 Phosphate Dehydrogenase deficiency (G6PD-d) in Malaria Endemic South Central Timor, East Nusa Tenggara, Eastern Indonesia

Jontari Hutagalung, M. Soleha, Nikson Sitorus and Linawati Hananta

Abstract

Approximately 18 million people live in malaria-endemic areas with 218,450 reported confirmed cases and 161 reported deaths in Indonesia. Currently, primaquine (PQ), the 8-aminoquinolines, is still the only drug for radical cure and preventing relapse of malaria. However, the individuals with G6PD deficiency (G6PD-d) have risk of hemolysis. Currently, few data of the prevalence of G6PD-d and genotyping are available. This study will provide the prevalence of G6PD-d and the genotyping in malaria cases in South Central Timor (TTS) district. G6PD status was analyzed with quantitative (Randox G6PD test, UK) follow with PCR-RFLP and sequencing to identify the variant of G6PD-d genotyping. Malaria was confirmed by n-PCR (Promega, Madison, USA). A total 64 of 181 individuals with G6PD-d from South Central Timor (TTS) district were analyzed. About 25 of 64 cases of G6PD-d were tested positive for malaria with *P. vivax* as the dominant species 56% (14/25) and most of the cases were female 73.3% (11/15). Among the 64 G6PD-d the genotyping *Vanua Lava* (10,883 T>C) WHO classifies G6PD-d genetic variants class II with severe deficiency <10% the enzyme activity were dominant. The variant of *Vanua Lava* is dominant and the high G6PD-d indicated that screening for G6PD deficiency is necessary.

Keywords: G6PD-d, endemic malaria, Eastern Indonesia

1. Introduction

Glucose-6-phosphate dehydrogenase deficiency (G6PD-d) is an X-linked genetic disorder that impacts insufficient enzyme activity. Moreover, 140 allelic variants of G6PD-d mutations are known and has been published [1]. The World Health Organization (WHO) divided G6PD-d into five classes according to the level of enzyme activity in the red blood cells (RBC), and the clinical manifestations [2].

IntechOpen

The importance from the variant of genotyping G6PD-d particularly for screening the risk of haemolytic anaemia is induced by the antimalarial primaquine (PQ) although the relationship between G6PD-d with protective to malaria infection is still unknown [3].

The gene of G6PD is an X-linked recessive hereditary disease characterized by abnormally low levels of glucose-6-phosphate dehydrogenase, a metabolic enzyme involved in the pentose phosphate pathway, especially important in red blood cells (RBC) metabolism [4]. Haemolytic anaemia in G6PD-d can be triggered by a range of oxidative agents, such as infections and certain foods and drugs, including antimalarial primaquine [5, 6].

During 2016–2017, the WHO reported that more than 16.7 million population in Indonesia were still living in high malaria transmission areas and the epidemiological distribution of malaria confirmed cases were 218,450 with 161 deaths cases reported. The identification was by *Plasmodium* species: *P. falciparum* 62% and *P. vivax* 37%. Up to date in Indonesia, the only one available antimalarial treatment for radical cure is PQ with 0.25 mg/kg, unfortunately with unknown variant genotyping of G6PD-d [5]. However, a limited study data on the prevalence and variant genotyping of G6PD-d are available in Indonesia, especially in malaria endemic areas. Tantular et al. 2010 studied the genotyping of G6PD-d among students of 7–12 years old in Maumere district, Flores Island, and found that the prevalence of G6PD-d was 4.4% (16/363) and also found there were five different variant genotyping of G6PD gene mutation although in a small population: 31% (5/16) *Coimbra* 592 C>T, and 18.7% (3/16) *Kaiping* 1388 G>A; other case variants are *Vanua Lava* 383 T>C, *Viangchan* 871 G>A, and *Chatam* 1003 G>A, each one cases [4]. Satyagraha et al. 2015 studied that G6PD-d in Sumba District, Eastern Indonesia was 5.1% (104/2033) with the mean average; the activity of G6PD enzyme was <4.6 U/gHb, normal 10 U/gHb; however, in contradiction with others' studies, the dominant genotypes were 98.5% (69/70) variant *Vanua Lava* 383 T>C [3]. Because G6PD-d caused chronic non-spherocytic haemolytic anaemia if PQ is consumed, and for this reason, the variant genotyping must be can identifiable in the endemic malaria areas; moreover, most of the G6PD-d are asymptomatic [7].

This study determines the prevalence of G6PD-d and the genotypes of G6PD mutation variants of malaria cases in malaria endemic, Eastern Indonesia, in part, upon previous reviews of variant G6PD-d genotyping by authors, and these may be consulted for more detailed information.

2. Methods

This study was undertaken at five sites (districts) with different levels of malaria endemic in East Nusa Tenggara Province (NTT), Eastern Indonesia. Ethic Approval for this study was approved by ethical committee of the Faculty of Medicine, Universitas Gadjah Mada, Yogyakarta, Indonesia, with reference no: KE/FK/85/EC. Data were collected from August 2014 to September 2014, and the whole blood samples were taken from the previous study of Hutagalung et al. [8, 9]. The criteria of inclusion were positive malaria from thick and thin blood smears by Giemsa 3–5% and nested PCR [10], more than 14 years of age, and signed a written informed consent. The study areas were selected by Annual Parasites Incidence (API) based on Health office data 2013/2014 [4]. All participants were interviewed face to face with standard questionnaire, and before enrolment study, all the participants had a physical examination by local health practitioners.

A total of 64 samples diagnosed with G6PD-d were defined by a quantitative G6PD-d test calculated and read at 340 nm/min with a spectrophotometer according to the manufacturers' protocols from the Randox G6PD test, UK, cat no: PD-410; normal G6PD is <6 U/gHb. The calculation of G6PD activity is as follows: G6PD mU/grHb = mU eritrosit per mL × 100/Hb g/dL. All the assessments of G6PD-d were observed with the G6PD normal control [11]. The assessment of the activity G6PD enzyme is as follows:

The DNA for G6PD genotyping was extracted from the frozen (−20°C) EDTA blood samples and the PCR using the Promega, USA (cat no. A−1120) [12], DNA Mini Kit manufactories protocols. G6PD gene PCR amplification and variant detection using five sets specific primers (25 mM) only for G6PD-d variant from exon 5, 6, 9, 11 and 12 because the reason most common variants in Asia Nguyen et al. [13]. The total volume of PCR G6PD gene was performed in 30 μL total reaction (15 μL green master mix PCR, 1 μL each forward and reverse primer, 2 μL DNA template, and 11 μL nuclease free water/ddH_2O). PCR temperature condition was followed by time melting (Tm) calculation. Positive and negative controls from known samples. The electrophoresis using 100 V and 55–60 minutes was run on agarose 1.5–2% from the Bioron-604001, Germany, containing 5 μL ethidium bromide, Promega, Madison, USA, cat. no. H-5041. The DNA ladder is of (2–3 μL) 100 bp, from the Vivantis, Selangor, Malaysia, cat. no. NL-1407. Finally to identified of genotyping of variant G6PD-d 25 μL PCR product for sequencing to Macrogen Laboratories and all the sequence of each samples was compared to the GenBank accession reference no. X-554481 [6]. The study flow was explained in **Figure 1** and **Table 1**.

While the identification of the *Plasmodium* was using DNA isolation from the whole blood samples, which were collected from each participant, transferred in EDTA anticoagulant tube BD Vacutainer 5 mL, the identification of *Plasmodium* using double assignment microscopic test by Giemsa 3–5% followed nested PCR with five sets of primers (20 mM) with time melting PCR condition followed from Snounow et al. [10, 12].

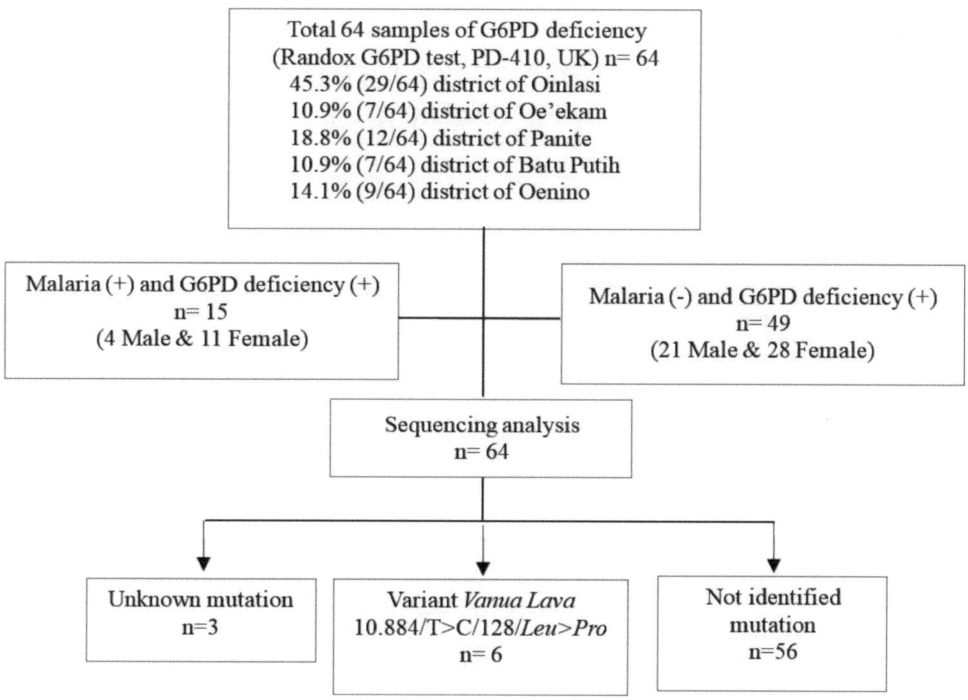

Figure 1.
The study flow.

Results of G6PD (U/gHb)	(+) Malaria/%	(−) Malaria/%	Total/%
Severe deficient 0–3	6/24	19/76	25/39
Moderate deficient 4–6	9/8.8	30/11	39/61
Total	15/23.4	49/76.6	64/100

Table 1.
Results of G6PD quantitative test (n = 64), G6PD deficiency test from Randox G6PD test, UK, cat no: PD-410.

3. Results

A total of 64 samples were detected; G6PD-d samples from five district study areas in South Central Timor District, East Nusa Tenggara Province, the Eastern Indonesia were included (**Figure 2** and **Table 2**), in which the mean of age was 42 years old (16–80 years old). During the study, 8.3% (15/181) samples with G6PD-d were infected with malaria. The result of this study showed that the mean average of the activity of G6PD enzyme of the samples was <10.4 U/gHb. The haemoglobin were no different between G6PD-d and normal 13.4 gr/dL (11–17 gr/dL). Almost all the samples were original from Timor ethnicity 99.8% (63/64). The prevalence of malaria in G6PD-d was 8.3%, in which *P. vivax* was the dominant species. However, this study results that the mix infection of malaria in G6PD-d were the *P. falcifarum* and the *P. vivax* infection which was 16.6% (3/15).

This map presents the distribution of the samples were G6PD-d in eastern Indonesia follows the malaria cases. The biggest cases were found in Oinlasi district 45.3% (29/64). However, all the sub-district also present the G6PD-d cases.

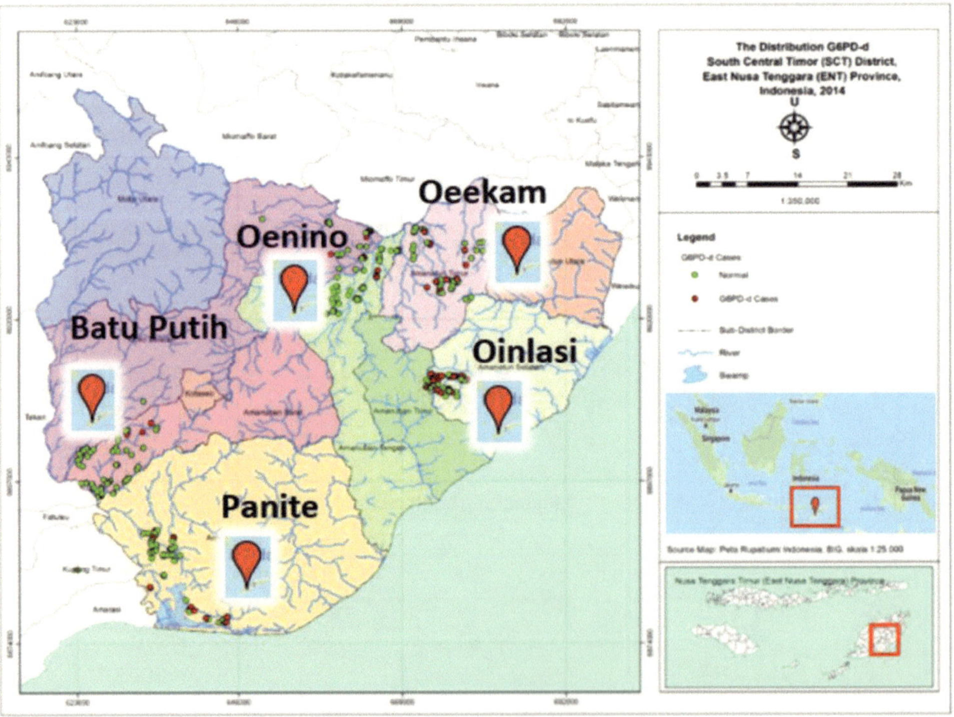

Figure 2.
Map of the five study areas and the distribution of the samples in South Central Timor District, East Nusa Tenggara, Province (Prov. NTT), Eastern Indonesia.

Mutation	Variant	Location	Amino acid	Enzyme activity
10.884 T>C	*Vanua lava*	Exon-5	128 Leu > Pro	<10% normal
—	—	Exon-6		
13.153 T>C	Unknown	Exon-9	372 Ser > Pro	<10% normal
—	—	Exon-11		
—	—	Exon-12		

Table 2.
The G6PD genotypes in South Central Timor District, East Nusa Tenggara Province, Eastern Indonesia. GenBank accession reference no. X-554481 [6].

Across the samples of G6PD-d, this study found that the significant 39% (25/64) was severe G6PD-d with the result of quantitative test <3 U/gHb. However, in line with this study, were also it was found that 23.4% (15/64) G6PD-d was due to malaria infection.

Totally, we analysed 64 samples with positive G6PD-d from five districts in Eastern Indonesia randomly. Among 15 samples with G6PD-d positive malaria dominantly were female 73.3% (11/15); these results in line with 76.5% (49/64) of the G6PD-d were also female 57.1% (28/49).

From the sequence analysis result most of the sequence we can not identified the mutation of the samples only 9.4% (6/64) identified with variant G6PD-d *Vanua Lava* (10.884 T>C) amino acid substitution and we also founded 4.6% (3/64) with unknown mutation. G6PD genotypes dominant were *Vanua Lava* (10.884 T>C) in exon 5 with amino acid substitution and unknown mutation 13.153 T>C in exon 9, respectively.

4. Discussion

This study findings 9.4% G6PD-d prevalence with variant Vanua Lava 10.884 T>C were dominant (WHO classifies in to class II with severe deficiency <10% the enzyme activity). This study result relevant with the previous study of G6PD-d prevalence in Asia among malaria patients from public health centre in Myanmar was 19.8% (50/252), and at the endemic malaria population areas in Sri Lanka 10.9% (225/2059), showing the prevalence; however, very different to the result of variant genotyping for G6PD-d the most dominant are variant *Mahidol* 487 G>A and *Kaiping* 1388 G>A, the varied in of G6PD-d we notion caused of the different of region and the ethnic status [14, 15]. Unfortunately, this study resulted with the prevalence more higher than the previous study in the same island and showed that the prevalence of G6PD-d was 5.9% (104/2033); however, this study still significant consisted with the variant genotyping were *Vanua Lava* 10.884 T>C was dominant, caused the same of region and ethnicity [3].

This result imposes a greater complexity to consider such antimalarial PQ therapy as more complicated. From the postulated evolution of G6PD-d, it is possible that the frequency of malaria is lower in patient with G6PD-d, though a protective effect against [14]. This consisted of the G6PD-d lower prevalence 7.8% (15/191) vs. non-deficiency 25.6% (49/191). However, from the previous study, in Indonesia, it was showed that there was significance between G6PD-d with malaria infection $p \leq 0.001$ [3, 16]. We would conclude from the theory and previous study that the individual with G6PD-d would perhaps more prove vulnerable and very sensitive to antimalarial PQ therapy in terms of risk of anaemia haemolytic and the variant genotyping [15, 17]. Authors suggest that G6PD-d induced haemolytic anaemia within 14 days after drug administration of PQ; to prevent this side effect, screening of G6PD-d is necessary prior to drug administration of PQ in malaria patients. Indeed, screening for G6PD-d prior to PQ therapy would likely must to be required to protect patients diagnosed especially with *P. vivax* infection [12, 17]. High G6PD-d were found in endemic malaria areas from public health perspective it is remains screening of G6PD-d testing must have to be taken, because in Eastern Indonesia until currently the combination of the PQ with ACT as the only remains for the first line antimalarial drug for radical cure hypnozoites and control for malaria transmission [18, 10, 19, 20]. Indeed, the risk associated with its use must be minimized during reach elimination phases [11, 9]. Patient with malaria should be tested for the enzyme activities and adequately be informed before administration of PQ after knowing the G6PD-d status of the patient, which is a pre-requisite for prescribing PQ with lower doses for individual. The WHO suggests using an intermittent PQ regimen of 0.75 mg base/kg once a week for 8 weeks to help malaria pre-elimination program smoothly [3, 21].

From this study, the prevalence of malaria based on the nested PCR result 32.6% was found with the *P. vivax* and *P. falciparum* dominants. Malaria is a major public health burden in Eastern Indonesia. High asymptomatic infection revealed in this study implied with the fact that asymptomatic malaria is common in high malarious areas and highlighted that low parasitemia or asymptomatic cases should be identified during implementation of the malaria elimination program [22, 23, 24]. This study suggests the active and passive case findings coupled with periodic mass blood surveys (MBS), case management with effective drugs, vector control and good surveillance more needed [25, 10]. This study also agreed about the prevalence of malaria by n-PCR resulted from the previous study from around in malaria endemic areas in Indonesia, Papua from 2004 to 2013 the study showing the mean average of the prevalence above 5–29.7% and higher resulted in Papua, Indonesia 53.3% (86.799/162.966) and more higher in children vs. adult people [23, 19]. This study is also relevant with the study measurement of the prevalence of malaria in other studies in Eastern Indonesia; but, it was still found that the prevalence was 46.4% (86.797/186.869). This study suggested that malaria infection associated with haematological impact diseases, such anaemia and morbidity also greater mortality caused malaria, therefore the public health very importance to plan how to control strategies in areas were the malaria infection was high [26, 27].

Mapping results almost all for the population at risk show that G6PD-d cases follow with the malaria distribution cases. The malaria risk displaying whole area study site, here additionally advantage this maps provide the base for design of the surveillance strategy and will be fully implemented by targeted inland hotspot for prevent outbreak occurred and endemicity maps were used to estimate real incidence malaria and G6PD-d in areas for pre-elimination began [28, 14]. In summary, our survey of G6PD-d in Eastern Indonesia the variant of *Vanua Lava* (10.884 T>C) was relative common among Asia and Eastern Indonesia. The significant risk for increase of the haemolytic cases after the PQ treatment has potential to induce oxidative stress. G6PD-d assessment should be done before antimalarial drug administration.

5. Conclusion and recommendation

The Prevalence of G6PD-d in South Central Timor District was 11.5% with variant G6PD-d *Vanua Lava* (WHO class II) dominant, means required screen before giving the administration of PQ and further research is needed to identified samples with unknown mutations.

Acknowledgements

Many thanks go to all the respondents in this study in of SCT District, ENT Province, Eastern Indonesia. This study was funded by Asia Pacific Malaria Elimination Network (APMEN), project number 103-02 as part of the APMEN Country Partner Technical Development Program, Round X. Major funding for APMEN is provided by Australian Government Department of Foreign Affair and Trade, alongside funds received from the Bill and Melinda Gates Foundation. Our special thanks go to the National Institute of Health Research and Development (NIHRD), Ministry of Health (MoH), Republic of Indonesia, Center for Tropical Medicine (Tropmed), Faculty of Medicine, Universitas Gadjah Mada. Province & District Health Office (PHO & DHO), and local laboratories, East Nusa Tenggara (ENT) Province. Eijkman Institute for Molecular Biology (Malaria I, II and RBC & Enzyme disorder Laboratory). We are deeply indebted to Prof. Rosichon, Dr. Satya Nugraha, Prof. Mo Salman (College of Veterinary Medicine and Biomedical Science, Colorado State University, Fort Collins, USA), Dr. Alden Henderson (Centers for Disease Control and Prevention, USA), Dr. Ari Winasti Satyagraha and Dr. Rintis Novijanti (Eijkman Institute for Molecular Biology, Jakarta) for review and editing this manuscript, and lastly, we wish to thank Mr. Alex H. Rambadeta as a freelance editor for his assistance in advising on the grammatical and spelling accuracy of the text where appropriate.

Author details

Jontari Hutagalung[1*], M. Soleha[1], Nikson Sitorus[2] and Linawati Hananta[3]

1 Centre for Biomedical and Basic Health Technology, The National Institute of Health Research and Development (NIHRD), Ministry of Health (MoH), Republic of Indonesia

2 Centre for Research and Development of Public Health Efforts, The National Institute of Health Research and Development (NIHRD), Ministry of Health (MoH), Republic of Indonesia

3 Department of Pharmacology and Pharmacy, School of Medicine and Health Sciences, Atma Jaya Catholic University of Indonesia, Jakarta, Indonesia

*Address all correspondence to: jontarihutagalung@yahoo.com

References

[1] Amoah LE, Opong A, Torgby RA, Abankwa J, Acquah FK. Prevalence of G6PD deficiency and *Plasmodium falciparum* parasites in asymptomatic school children living in southern Ghana. Malaria Journal. 2016:4-11

[2] Matsuoka H et al. Five different glucose-6-phophate dehydrogenase (G6PD) variants found among 11 G6PD-deficient persons in Flores Island, Indonesia. Journal of Human Genetics. 2003;**48**(10):541-544

[3] Satyagraha AW et al. G6PD deficiency at Sumba in eastern Indonesia is prevalent, diverse and severe: Implications for Primaquine therapy against relapsing Vivax malaria. PLoS Neglected Tropical Diseases. 2015;**9**(3):1-17

[4] Tantular IS et al. Incidence and mutation analysis of glucose-6-phosphate dehydrogenase deficiency in eastern Indonesian populations. Acta Medica Okayama. 2010;**64**(6):367-373

[5] WHO South East Asia Region. Indonesia Malaria Country Profile; 2016

[6] Minucci A, Moradkhani K, Hwang MJ, Zuppi C, Giardina B, Capoluongo E. Glucose-6-phosphate dehydrogenase (G6PD) mutations database: Review of the 'old' and update of the new mutations. Blood Cells, Molecules & Diseases. 2012;**48**(3):154-165

[7] Saraswathy KN, Sachdeva MP. G6PD deficiency, sickle cell trait, haptoglobin and transferrin polymorphisms among Koyadoras and Nayakpods of Andhra Pradesh. Anthropologist. 2008;**10**(2):163-165

[8] Hutagalung J, Kusnanto H, Supargiyono, Hamim SA, Satyagraha AW, Novijanti R, et al. Outbreak Surveillance and Investigation Reports (OSIR). Malaria Pre-elimination Assessment in Eastern Indonesia. 2016;**9**(1):1-7

[9] Hananta L, Astuti I, Sadewa AH, Alice J, Hutagalung J, Mustofa. The prevalence of CYP2B6 gene polymorphisms in malaria-endemic population of Timor in East Nusa Tenggara Indonesia. Osong Public Health and Research Perspectives. 2018;**9**(4):192-196

[10] Murhandarwati EEH et al. Change of strategy is required for malaria elimination: A case study in Purworejo District, Central Java Province, Indonesia. Malaria Journal. 2015;**14**(1):1-14

[11] Ghani RA et al. Glucose 6-phosphate dehydrogenase deficiency among Yemeni children residing in malaria—Endemic areas of Hodeidah governorate and evaluation of a rapid diagnostic test for its detection. Malaria Journal. 2016:4-13

[12] Seck MC et al. Malaria prevalence, prevention and treatment seeking practices among nomadic pastoralists in northern Senegal. Malaria Journal. 2017:1-11

[13] Hue NT et al. Glucose-6-phosphate dehydrogenase (G6PD) mutations and haemoglobinuria syndrome in the vietnamese population. Malaria Journal. 2009;**8**(1):1-8

[14] Phompradit P, Kuesap J, Chaijaroenkul W, Rueangweerayut R. Prevalence and distribution of glucose-6-phosphate dehydrogenase (G6PD) variants in Thai and Burmese populations in malaria endemic areas of Thailand. Malaria Journal. 2011;**10**(1):368

[15] Lee J et al. Prevalence of Glucose-6-Phosphate Dehydrogenase (G6PD) Deficiency Among Malaria Patients in Upper Myanmar2018. pp. 1-7

[16] Sari MI. Hubungan Defisiensi Glukosa 6 Fosfat Dehidrogenase (G6pd) Dengan Infeksi Malaria Pada Populasi Nias Di Kota Medan, 2008 USU Repository©2008. pp. 1-85

[17] Luzzatto L, Seneca E. G6PD deficiency: A classic example of pharmacogenetics with on-going clinical implications. 2014. pp. 469-480

[18] Howes RE et al. G6PD deficiency prevalence and estimates of affected populations in malaria endemic countries: A geostatistical model-based map. PLoS Medicine. 2012;**9**(11)

[19] Pusat Data dan Informasi and Direktorat Pengendalian Penyakit Bersumber Binatang. Epidemiologi Malaria di Indonesia. Buletin Jendela Data & Informasi Kesehatan. 2011;**1**:1-16

[20] Kementerian Kesehatan RI. Pedoman Manajemen Malaria2015. pp. 1-128

[21] Khim N et al. G6PD deficiency in *Plasmodium falciparum* and *Plasmodium vivax* malaria-infected Cambodian patients. Malaria Journal. 2013;**12**(1):0-9

[22] Tanner M et al. Malaria eradication and elimination: Views on how to translate a vision into reality. BMC Medicine. 2015;**13**(1):1-22

[23] Elyazar IRF et al. *Plasmodium vivax* malaria endemicity in Indonesia in 2010. PLoS One. 2012;**7**(5)

[24] Laishram DD et al. The complexities of malaria disease manifestations with a focus on asymptomatic malaria. Malaria Journal. 2012;**11**(1):29

[25] Smith DH et al. A sticky situatioon: The unexpected stability of malaria elimination. Philosophical Transactions of the Royal Society, B: Biological Sciences. 2013 (In press)

[26] Douglas NM et al. Unsupervised primaquine for the treatment of *Plasmodium vivax* malaria relapses in southern Papua: A hospital-based cohort study. 2017. pp. 1-19

[27] Sorontou Y, Asih PBS, Wanandi SI, Ramelan W, Syafruddin D. Malaria in Jayapura District, Papua Province, Indonesia and resistance to sulfadoxine-pyrimethamine. 2007. pp. 32-38

[28] Howes RE et al. Spatial distribution of G6PD deficiency variants across malaria-endemic regions. 2013. pp. 1-15

Chapter 8

Two Endemic Primates' Species in China: Hainan Gibbon and Guizhou Snub-Nosed Monkey

Jiang Zhou and Huaiqing Deng

Abstract

Hainan gibbon (*Nomascus hainanus*) is the most threatened species of gibbon, the endemic primates of China. Currently, there were only 4 groups no more than 30 individuals left, which is only distributed in a 16 km^2 area (H: 800–1280 m) of Bawangling National Nature Reserve of Hainan island in China. The social structure of the Hainan gibbon is polygynous, with one adult male pairing with two adult females. Hainan gibbon research has always been one of the hot spots of primate research, mainly concentrated in ecology, behavior, and genetic research. Here, we mainly reported the classification, historical population changes, community structure, dietary, reproductive and song behavior of Hainan gibbon. We also reported the other unique primate of China: Guizhou snub-nosed monkey.

Keywords: primate, species, China, Hainan gibbon, Guizhou snub-nosed monkey classification

1. Introduction

Hainan gibbon (*Nomascus hainanus*) is the most threatened species of gibbon, the endemic primates of China. Currently, there were only 4 groups no more than 30 individuals left, which is only distributed in in a 16 km^2 area (H: 800–1280 m) of Bawangling National Nature Reserve of Hainan island in China [1]. The social structure of the Hainan gibbon is polygynous, with one adult male pairing with two adult females [2]. Hainan gibbon research has always been one of the hot spots of primate research, mainly concentrated in ecology, behavior, and genetic research.

Guizhou snub-nosed monkey (*Rhinopithecus brelichi*) belong to the primates, Cercopithecidae, Colobinae, *Rhinopithecus*, is one of the most endangered (En) species and first-class protected animals in China, according to the International Union for the Conservation of Nature and Natural Resources [3]. These monkeys reside only at the Fanjingshan National Nature Reserve (FNNR) in a narrow habitat that can fragment severely under risky survival conditions [4]. Similar to other Chinese snub-nosed monkey species, habitat destruction is a potential threat to the survival of the Guizhou snub-nosed monkey [5, 6], and suitable habitat at low altitudes has come under increased pressure in the FNNR. Due to the development of tourism at both FNNR and Mt Fanjingshan, the habitat of the monkey has become severely disturbed.

We summarized these two primates species research results in recent years as follows.

2. Methodology

2.1 Literature collection

By searching for the historical documents, to understand the Hainan gibbon name origin and classification changes. Hainan gibbon and Guizhou snub-nosed monkey ecological characteristics such as population change, population structure and distribution.

2.2 Wild survey

We have 18 years of wild survey for Hainan gibbon in Bawangling nature reserve since 2002. We also interviewed residents to understand the history of Hainan gibbon distribution and the current status. Hainan gibbon has the characteristics of aloud singing in the morning. So, we can determine the location of Hainan gibbon according to singing and track observation. We used a recorder to record the Hainan gibbons' songs behavior and alarm calls. We also observed the reproductive behavior, feeding behavior and other behaviors. To understand the Hainan gibbon habitat for plants, we set up 85 quadrants (30 × 30 m) of plants according to the elevation gradient at an altitude of 600–1200 m in Bawangling nature reserve.

Guizhou snub-nosed monkey is only distributed in Fanjingshan National Nature Reserve, Guizhou, China. The area of this reserve is 567 km^2 (27°46′50″–28°1′30″N, 108°45′55″–108°48′30″E). We used telescope to observe the behavior of monkeys, such as feeding behavior, breeding behavior, resting and playing behavior. At the same time, we used GPS to record the location of behavior. We collected the plant specimens of the monkeys feeding for species identification, and record the part of the plant was consumed by the monkeys.

To understand the habitat conditions of Guizhou snub-nosed monkey, we set up 100 botanical plots at different altitude in the Fanjingshan National Nature Reserve. The area of each plot was 500 m^2. The plant species name and number of the botanical plots were recorded. The species of relative base coverage, relative frequency and relative density were analyzed.

3. Results

3.1 Species description of the two primates

3.1.1 Classification of the Hainan gibbon

Hainan gibbon (*Nomascus hainanus*) (**Figure 1**) is the earliest named for the first time as a new species of *Hylobates hainanus* by Thomas [7] according to a specimen of the Hainan gibbon stored in the British natural history museum. Since then, the classification status of the Hainan gibbon has gone through several changes; first Pocock [8] put it as the *Hylobates nasutus* species; then Delacour [9] argued that the Hainan is supposed to be a subspecies of black gibbons (*H. naustus*) from Tonkin regions, and named *H. concolor hainanus*. But later, researchers found that they are significantly different in shape, color and other characteristics between them; so, the Hainan gibbon is taken as an independent subspecies [10–14]. Chinese scholars

Figure 1.
The adult female of the Hainan gibbon with infant.

Su Bing et al. [15] by molecular genetic evidence proved the Hainan gibbon and other black gibbons have bigger difference in molecular level , and it should be an independent species. Geissmann [16] classified the Hainan gibbon as Hainan subspecies (*Nomascus sp. cf. Nasutus hainanus*) of East black gibbons, according to the differences in color and singing behavior; after that Brandon-Jones et al. [17] proved Hainan gibbons and East black gibbon are supposed to be two different subspecies by the evidence of molecular biology. Recent research results show the Hainan gibbon is an independent gibbon species based on morphology and singing behavior [18–21].

3.1.2 Morphological features of Rhinopithecus brelichi

In 1903, the British biologist Thomas visited Fanjingshan and named the Fanjingshan snub-nosed monkey as Guizhou snub-nosed monkey (*Rhinopithecus brelichi*) (**Figure 2**).

Guizhou snub-nosed monkey is smaller as Sichuan snub-nosed monkey morphologically, but the tail is longer than Sichuan snub-nosed monkey. Body length of Guizhou snub-nosed monkey is 637–690 mm and tail length is 846–905 mm. The

Figure 2.
Wild Guizhou snub-nosed monkey in Fanjingshan National Nature reserve.

nose of the monkeys is slightly concave downward. The color of face is gray or pale blue. The body of Guizhou snub-nosed monkey is gray in color, with clear white patch between the shoulders. From the shoulders down the upper lateral to the back of hand, the color changes from light gray gradient in brown to black. Under the neck, armpit and upper limb inside are gold, sharing yellowish gray color. Tail base is dark gray, to the end changes to black or yellowish-white, black around the callosity. Juveniles color were pale, perfect silver gray. The tail was dark gray end of yellow-white.

3.2 Historical population of the two primates

3.2.1 Historical population changes in Hainan gibbons

Hainan gibbons were widely distributed in Hainan province until the 1960s. In the early 1950 s, there were still about 2000 individuals of Hainan Gibbons in Hainan Island. In 1978, the Hainan gibbon has narrowed the range of the Baisha County, Qiongzhong County and Limushan Mountain, population has dropped to only 450–500 individuals. But by the early 1990s, they were distributed only in Bawangling National Nature Reserve. The other sub-populations were extirpated during a period of logging and hunting between 1960 and 1980. The current population increased from a total of 10 individuals in 2–3 groups in 1978 to 21 individuals in 4 groups in 1988 [2]. The population decreased to 15 in 1993 as a result of hunting [22]. There were only 13 individuals remaining in 2002, including 6 in group A (including 2 infants), 5 in group B (including 1 infant) and 2 solitary individuals. Population changes in recent years is shown in **Figure 3**. The rate of population recovery has increased progressively since then and increasing up to 26 in 2013 [23]. A new family group C is formed in the summer of 2011, later the group D is formed in 2016. As a result of male and female sex ratio, the female is less, so the newly formed group D has only one adult female. An infant was born in group C in 2018, now there were 4 groups of Hainan gibbon and 28 individuals distributed in Bawangling National Nature Reserve.

3.2.2 Historical population and distribution changes of Guizhou snub-nosed monkeys

Quan and Xie [24] first reported that Guizhou snub-nosed monkey population is about 200 individuals, Tan [25] conducted preliminary investigation, and estimated its population to be 300–500. Eudey [26] reported that Guizhou snub-nosed

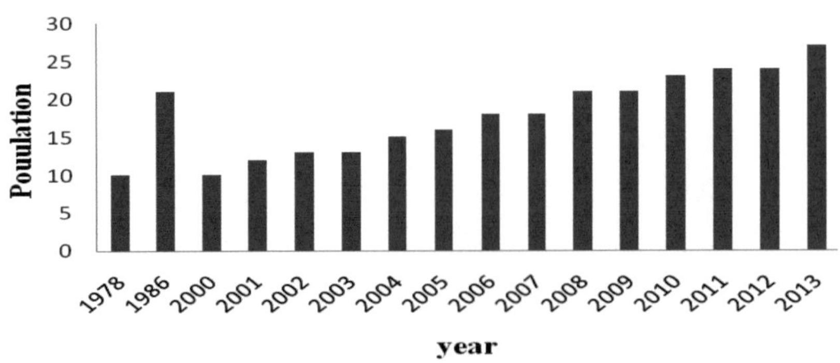

Figure 3.
The changing number of the Hainan gibbon population in the Bawangling natural reserve.

monkeys do not exceed the total number of 500–670. In 1993, the Fanjingshan National Nature Reserve reported that the population of Guizhou snub-nosed monkeys was estimated at 764 individuals [27]. Bleisch [28] reported there are 800 individuals, Quan and Xie [29] think the population is about 500 individuals. Yang et al. [27] identified the wild populations as 750–764 after years of investigation. Then the survey indicates that the population size is approximately 750 monkeys [30]. A recent population survey shows that the Guizhou snub-nosed monkeys is 661–710, belonging to 94 "family groups" [31], almost unchanged from 1993. The *Rhinopithecus brelichi* acts as a group structure, mainly active in four regions in the Fanjingshan National Nature Reserve. Each group consists of 2–10 family groups with the smallest family size consisting of 1 male and 1 female. The common family group consists of 1 adult male and 2–4 adult female individuals, and sub-adult individuals, young individuals and infants. Compared with *R. bieti* and *R. roxellana*, the Guizhou snub-nosed monkeys have a lower diversity of mitochondria DNA, which is vulnerable to environmental changes or climate oscillations [32–34].

However, the range of habitats available to the Guizhou snub-nosed monkey has changed dramatically. In the 1960s, Guizhou snub-nosed monkeys were often observed at an elevation of 500 m [29]. In the 1980s, they were commonly observed at 800 m and above [29]. Studies conducted in 1990s and between 2000 and 2009 showed an elevation range 1400–2100 m [4, 29, 30]. Now they are distributed in the nature reserve with the altitude 900–2180 m [31].

3.2.3 Artificial breeding of Guizhou snub-nosed monkey

In 1992, the state forestry administration approved Fanjingshan national nature reserve captured seven individuals from wild to conduct breeding research in view of Guizhou snub-nosed monkey fewer wild populations, small range of habitats and particularly high extinction coefficient. After 3 years of hard work, the first artificial breeding of Guizhou snub-nosed monkey was born in 1995. In 2005, the scientific research personnel of Fanjingshan national nature reserve administration succeeded in breeding the second generation of Guizhou snub-nosed monkey.

Generally speaking, the wild animals captive breeding reach the second generation and reach a certain amount, the animal's growth is good, it marked the artificial breeding technology is complete maturity. But by the end of 2015, Fanjingshan national nature reserve administration only had five individuals of Guizhou snub-nosed monkey. In 2007, Fanjingshan national nature reserve gave one male and one female captivity Guizhou snub-nosed monkey to the state forestry administration in Beijing Daxing wildlife rescue center, later transferred to Beijing Zoo breeding, just one infant was born in 2013. The captive populations of Guizhou snub-nosed monkey also have a lot of problems in terms of breeding.

3.3 Ecology research of the two primates

3.3.1 Hainan gibbon community structure

Gibbon's main social organization is monogamy, the average group size of four individuals [35], but in the black gibbon is polygamy [36]. While Hainan gibbon group size is commonly four to nine individuals, each group including one adult male, two adult females and their offspring [2, 37]. Jiang et al. [38] research results show that the two groups of Hainan Gibbon, which is distributed in Bawangling National Nature Reserve gibbon began in 1986, each family group has an adult male and two adult females. In 1994, a new family group formed its social structure is monogamy, but the original two family group still has one adult male and two adult females. By 1997, the newly formed family group becomes the second wife of the social structure, protection zones of all three group of gibbon to form the two adult females social structure. During 2002–2015, when we are doing research in Bawangling National Nature Reserve of Hainan province, we found that one newly family group was formed in 2011, the new group social structure also has one adult male and two adult females. We also found in the process of investigation and research of Hainan gibbon, the biggest extended family group size is nine individuals; when the males of family group are about to reach sexual maturity, they are automatically expelled by adult male of family group leaving birth group, generally in independent living surrounding the family unit.

3.3.2 Habitat and the home range of Hainan gibbon

Chivers [39] reported that the gibbons generally live in tropical lowland rainforest or subtropical mountain evergreen broad-leaved forest, occasionally some species will be distributed in evergreen forest and half deciduous forest near the northern mixed environment. Hainan gibbon research started relatively late, the first habitat research was completed by Liu et al. [40], in their study they showed that in Bawangling nature reserve, the Hainan gibbon's general activities were at an altitude of 800–1200 m, home range area is larger.

The early results of Hainan gibbon's home range is 3.05–5.0 km^2 [41], the later scholars used GIS system to calculate the home range is 5.48–9.87 km^2 [42, 43]. Hainan gibbon's home range is far greater than other species of gibbon family groups, such as living in Thailand region of white palm gibbon and Java moloch gibbons domain area of only 0.16 and 0.17 km^2, life in Bangladesh hoolock gibbons and Malaya area of lar gibbons has larger home range area of only 0.45–0.56 km^2 [44].

3.3.3 Temporal habitat utilization of Guizhou snub-nosed monkey

To investigate the vegetation distribution of Guizhou snub-nosed monkey habitat, botanical plots were set up at 1500, 1700, 1900, and 2100 m in the FNNR. This is the altitudinal range the monkeys exploited. Each plot was 50 m (altitude) × 10 m (width) or 500 m^2. Plot size was based on the steeply sloped nature of the landscape of the Fanjingshan Mountain. The total number of botanical plots constructed and monitored was 100. We identified 236 species from 125 genera and 72 families in the 100 botanical plots, which included 183 tree species (75 genera and 42 families) and 53 vine species from 26 genera and 16 families. The Shannon-Wiener diversity and evenness indices for each vegetation community were higher at elevations 1700 m and 1900 m ($H = 5.97$ nit, $E = 0.85$, $H = 5.70$ nit, $E = 0.87$) and lower at elevations of 1500 m and 2100 m ($H = 5.24$ nit, $E = 0.82$; $H = 4.64$ nit, $E = 0.83$).

The exploitation of resources used by Guizhou snub-nosed monkey varied significantly by elevation among seasons (Kruskal-Wallis, $X^2 = 9.746$, $df = 3$, $p < 0.05$). From January to March (Spring), the elevational range (537–668 m) and daily distance travel (5200–7000 m) was greatest. In contrast during the summer and fall (April–September), the two seasons had the lowest elevational range of 144 m and the lowest daily distance travel of 550 m. From October to December (winter), the elevational range and daily distance travel gradually increased (**Table 1**).

3.3.4 Dietary of Hainan gibbons

We conducted dietary composition study of Hainan gibbons in Bawangling National Nature Reserve in China from August 2002 to August 2014. Total field observation days of all groups were up to 584, and accumulated 3400 h. The results

Month	Frequency of monkey activity	Mean elevation (m)	SD[a]	Minimum elevation (m)	Maximum elevation (m)	Elevational range(m)[b]	Maximum daily move distance(m)	Minimum daily move distance(m)	SD[c]
1	24	1730	210	1432	2100	668	6000	5400	147
2	25	1680	173	1560	2100	540	6200	5200	369
3	29	1799	161	1503	2040	537	7000	6100	242
4	31	1645	113	1510	1876	366	1800	1000	226
5	24	1764	109	1543	1887	344	1000	550	119
6	21	1812	44	1756	1900	144	2000	1200	229
7	18	1790	80	1654	1912	258	2000	1100	257
8	28	1757	106	1507	1805	298	3200	2600	160
9	29	1780	145	1524	1921	397	4000	3200	233
10	26	1824	156	1513	1979	466	4000	3500	146
11	31	1810	144	1590	2100	510	5200	4300	315
12	37	1773	140	1546	2000	454	6000	5100	274
Spring	78	1736	176	1432	2100	668	7000	5200	493
Summer	76	1744	122	1510	1900	390	2000	550	426
Fall	75	1764	121	1507	1921	414	4000	1100	820
Winter	94	1800	147	1513	2100	587	6000	3500	739

[a] *Standard deviation.*
[b] *Highest elevation minus lowest elevation in any month.*

Table 1.
Monthly variation in elevational range and maximum daily move distance, September 2011 to August 2013 [45].

show that Hainan gibbons consumed 133 plant species [46]. Of the plant species eaten, 81.1% were trees, 15.9% were lianas and 3.0% were herbs. Hainan gibbon spent 64.8% of feeding time on fruits, followed by on leaves (28.8%), buds (5.5%), flowers (0.6%) and animals feed (0.3%). The Hainan gibbon consumed fruits from 14 Ficus species, whereas it only accounts for 7.2% of the feeding time. The most commonly used part of a plant was ripe fruit in general. Five kinds of animal feed were eaten including young birds, bird eggs, spiders, termites and pupae.

Family	Species	Feeding part
Anacardiaceae	*Choerospondias axillaris*	Fr
Annonaceae	*Oncodostigma hainanense*	Fr
	Pouteria hainanense	Fr
	Polyalthia laui	Fr
Apocynaceae	*Melodinus suaveolens*	Fr
Araliaceae	*Schefflera octophylla*	Fr, L
Burseraceae	*Canarium album*	Fr
Elaeocarpaceae	*Elaeocarpus apiculatus*	Fr
	Elaeocarpus sylvestris	Fr
Euphorbiaceae	*Baccaurea ramiflora*	Fr
	Bischofia javanica	Fr
	Endospermum chinense	Fr, L
Guttiferae	*Garcinia multiflora*	Fr
	Garcinia oblongifolia	Fr
Lauraceae	*Litsea baviensis*	Fr
	Machilus chinensis	Fr
Moraceae	*Artocarpus styracifolius*	Fr
	Ficus esquiroliana	Fr
	Ficus harmandi	Fr
	Ficus tinctoria	Fr
	Ficus variegata	Fr
	Ficus virens	Fr, L
Myrtaceae	*Acmena acuminatissima*	Fr
Papilionaceae	*Derris trifoliata*	L
Polygalaceae	*Xanthophyllum hainanense*	L
Proteaceae	*Helicia kwangtungensis*	Fr
Rubiaceae	*Mussaenda hainanensis*	L
	Nephelium lappaceum	Fr
Tiliaceae	*Microcos paniculata*	Fr
Ulmaceae	*Gironniera subaequalis planch*	Fr
Vitaceae	*Tetrastigma cauliflorum*	Fr

Note: Part(s) eaten: L, leaves, Fr, fruit, Fl, flowers.

Table 2.
Plant species that can be used by Hainan gibbons per year at Bawangling National Nature Reserve.

The dominant plant families were Moraceae, Annonaceae, Lauraceae and Myrtaceae, with more than seven species. Hainan gibbons feed on two kinds of Leguminosae plants, *Derris trifoliata* and *Millettia pachyloba*, consuming their tender leaves during dry season, when other food is scarce. Plant family and species such as Moraceae, *Pouteria hainanense*, *Canarium album* and *Gnetum montanum* are the major food sources of Hainan gibbon, but during the dry season they bear only a few fruits.

There were only 32 food species repeatedly used by Hainan gibbons each year (**Table 2**), and 18.7% of species leaves were eaten [46]. There are few fig species available as foods especially in the dry season. In this season, due to less precipitation, Ficus fruit and non-figs become scarce and less available. Hainan Gibbons eat more tender leaves and increase their feeding time to live through this period.

3.3.5 Food resource distribution and availability of Guizhou snub-nosed monkey

The Guizhou snub-nosed monkeys were observed to feed on a total of 104 species from 51 genera and 31 families. These included 84 tree species from 41 genera and 26 families, representing 45.9% of the total number of tree species, and 20 vine species from 10 genera and 7 families, representing 37.7% of the total vine species. The IVI percentages of the feeding trees species at various elevations were: 65.5% at 1500 m, 65.6% at 1700 m, 72.9% at 1900 m and 58.7% at 2100 m. The IVI percentage of feeding trees accounted for 66.4% of the total number of records in the plots.

The plant species consumed by the Guizhou snub-nosed monkey varied significantly among seasons. (Kruskal-Wallis, $X^2 = 8.897$, df = 3, $p < 0.05$). From January to March, 34 feeding trees species accounting for 32.7% of the diet were consumed, mainly the buds of *Pterostyrax psilophyllus, Prunus grayana, Carpinus kweichowensis, Betula austrosinensis, Styrax japonica, Sassafras tzumu* and *Acer flabellatum*. From April to June, 64 feeding trees species accounting for 61.5% of the diet were consumed, mainly the leaves of *Tilia tuan, Prunus grayana, Styrax faber, Cyclobalanopsis gracilis, Cladrastis sinensi* and *Sorbus wilsoniana* and fruits of *Prunus grayana*. From July to September, 81 feeding trees species accounting for 77.9% of the diet were consumed, mainly the leaves of *Prunus grayana, Tilia, Litsea cubeba* and *Styrax japonica* and the fruits of *Dendrobenthamia and Acer davidii*. Finally, from October to December, 52 feeding trees species accounting for 50% of the diet were consumed, mainly the fruits and seeds of *Dendrobenthamia, Cerasus serrulata*, *Sorbus megalocarpa* and *Camellia* and the buds of *Magnolia sprengeri*. The monkeys fed on a core set of 28 species during all seasons the year, which accounted for 15.3% of the total number of species consumed. In addition, 48 species were consumed during at least 2 seasons of the year.

3.4 Behavioral ecology research of Hainan gibbon

3.4.1 Reproductive behavior

Hainan gibbon sexual maturity relatively late (generally around 7–8 years), generally produce one infant every 2 year [47]. The mating behavior of Hainan gibbon generally initiated by the female adult individuals, each mating duration is very short, and mate can range from 1 to 9 times. Posture is pushing type mating, pregnancy for 173 days. The infants can completely leave maternal independent activities that are needed 1.5 years after birth [1].

3.4.2 Song behavior

All gibbons have loud, complex, stably structured songs [48]. Gibbon songs occur frequently and regularly, usually beginning shortly before or after sunrise, with a usual duration of 10–30 min [49]. Gibbon song sound frequency varies among species, with a general fundamental frequency in the 0.2–5 kHz range [50], and male song reaching up to 4 kHz, such as in *Nomascus leucogenys* [51]. We recorded the songs of three Hainan gibbon groups and seven solitary Hainan gibbons in Bawangling National Nature Reserve, Hainan Province, China. In total, we recorded 184 segments of Hainan gibbon song over 129 days. Recordings from group B were conducted in 2002–2003 and the quality was too poor for analysis, so we concentrated on songs recorded in 2011–2013 from groups A and C.

The first song of Hainan gibbon started within 20 min (before or after) of sunrise, and the second 4 h after sunrise, during which time there were occasionally a third and fourth song event. We found that Hainan gibbons' snog includes female–male chorus and male solos, no female solos. A typical chorus is usually initiated and dominated by the adult male in Nomascus. Females add a great call and the chorus ends with male song [48, 51]. There was no significant difference in song duration between solitary gibbons and group C (single-factor analysis of variance and LSD multiple comparisons: F = 5.13, df 1 = 2, df 2 = 14, P = 0.26), but both differed significantly from group A (solitary: P2 = 0.016, group C = 0.008) [18].

3.4.2.1 Male song

Male song in groups was composed mainly of two types of sounds: simple repetition of one note and frequency-modulated (FM) notes. Simple repeated notes consisted of one to three similar, short notes. FM notes consisted of one to five long notes (**Figure 4**). Solitary adult male song sound spectrum characteristics were simple, consisting only of three to five long FM notes and no short or single notes. The long-note frequency of solitary gibbons was similar to that of group song, but the durations of each note and complete song were shorter than in the family groups.

3.4.2.2 Chorus

The Hainan gibbon chorus is usually initiated and dominated by an adult male. The females issue a simple great call at the climax of the male's song, and then the male contributes to the chorus with complex FM notes, and finally the chorus ends with a simple FM note sung by the adult male (**Figure 5**). Subadults and juveniles did not sing alone and were involved only in chorus song.

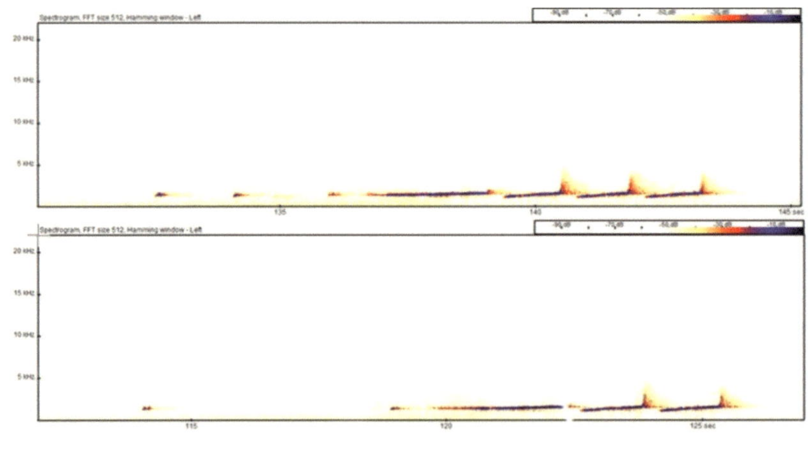

Figure 4.
Song spectra of Hainan gibbons in group A at Bawangling nature Reserve in September 2012 (Reference from [18]). (Note: The songs consist of two or three short syllables and three or four long syllables.)

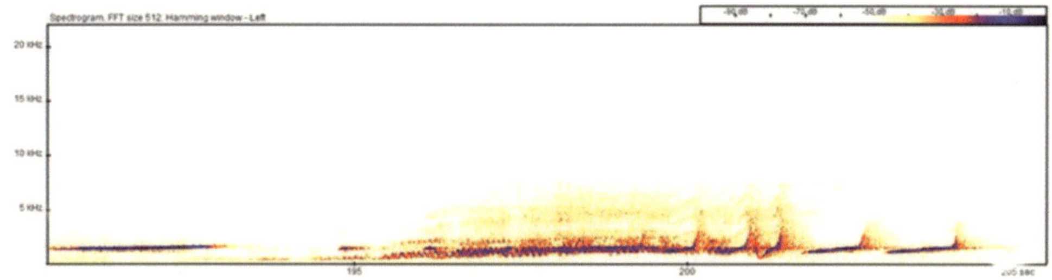

Figure 5.
Song spectra of duetting gibbons in group A at Bawangling nature Reserve in March 2013 (Reference from [18]).

In a full chorus, the male gibbon begins with a song consisting of two short notes and a long note: the male's song lasts 2–3 s. The start of the female's song is also a single, short note, followed by a loud call. The duration of choruses is ca. 7.5 s. Finally, the adult males sing an accompaniment, which also consists of very simple FM notes. In choruses, the song of male usually comprises two short notes and five to six relatively stable long notes. Compared with solo song, there is one additional long note. Male chorus and solo song did not differ significantly between groups A and C in frequency ($F = 1.417$, df 1 = 3, df 2 = 35, $P = 0.25$). Once the male Hainan gibbon has sung a FM note in chorus, it then sings songs that combine a short note with along FM note. After the first FM note, males continue with one to three short single notes and the sonic frequency shifts higher and higher. The next stage is loud song by two adult females, which then give two to four simple FM notes. The mean ± SD FM note duration was 5.20 ± 1.43 s (range 2.28–11.1 s); the starting frequency of the first FM note was 1.38 ± 0.06 kHz; it then rose to 1.64 ± 0.05 kHz (N = 59), with a duration of 2.67 ± 0.80s (N = 59); the starting frequency of the second FM note was 1.19 ± 0.06 kHz; it then rose to 1.65 ± 0.035 kHz, with a duration of 1.40 ± 0.03 s (N = 59). A complete female great call in chorus includes seven to nine notes. Female great call sound spectrum characteristics were very similar between groups A and C. Excepting the first two notes, the great call consists of rising fast trill notes. In the great call, the first note's duration is longer than that of subsequent notes (length 0.86 ± 0.40 s, N = 14) and its frequency is lower (0.62 ± 0.01 kHz, N = 14). The note of the great call then becomes shorter and the frequency increases until it reaches the highest frequency at the end. The frequency of the highest note made by adult females is at least 1.70 kHz, the highest frequency being no more than 2.0 kHz, with a mean of 1.78 ± 0.06 kHz (N = 13), the mean duration was 4.71 ± 1.23 s.

3.4.2.3 Alarm calls

Alarm calls are important defensive behaviors. We have reported the acoustic spectrum characteristics of alarm calls produced by Hainan gibbons (*Nomascus hainanus*) inhabiting Bawangling National Nature Reserve in Hainan, China [52].

Alarm calls of Hainan gibbon were emitted by family groups only and not solitary animals. Hainan gibbons produce two types of alarm call. Hainan gibbons emitted the same alarm call across different threats (raptor, snake or people). Male and female adult gibbons emitted "gou-gou-gou" to alert group members of a danger when they encountered a potential predator, hereafter termed an alarming call (**Figure 6**). Following the alarming call, individuals in the family group gather, young gibbons move rapidly towards females, adult males conceal themselves in a location convenient for further vigilance, and all members (except infants) produce "jier- jier-jier" sounds, occasionally accompanied by jumping in a tree or flapping

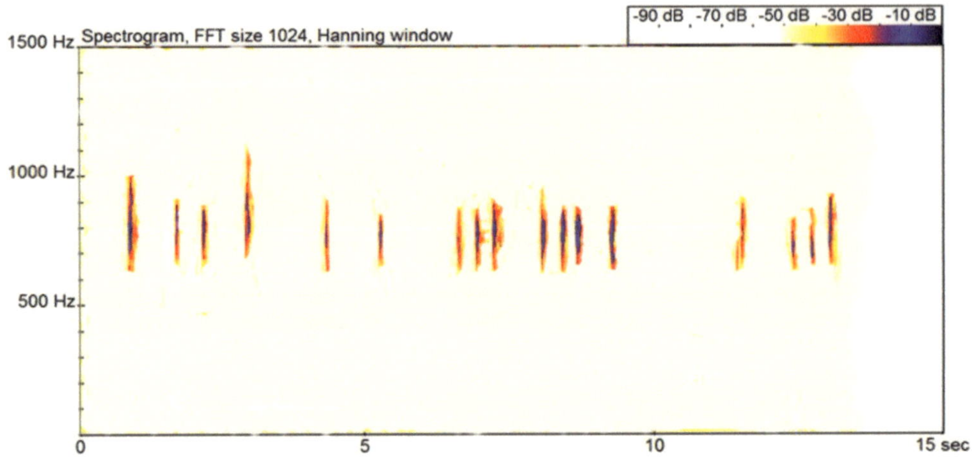

Figure 6.
The alarming call in a group of Hainan gibbons (group B) (Reference from [52]).

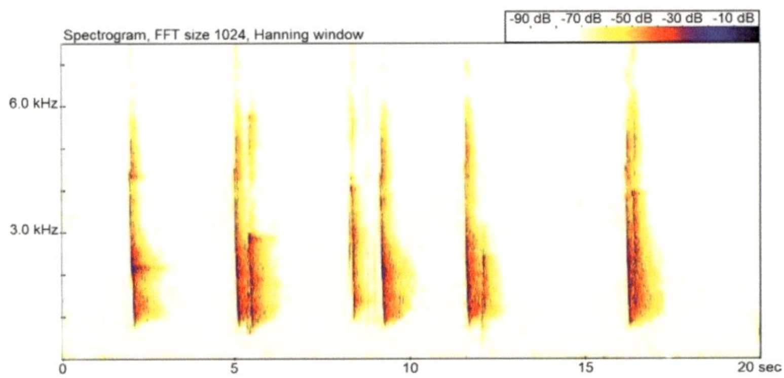

Figure 7.
The mobbing call in a group of Hainan gibbons (group B) (Reference from [52]).

tree branches; we term the group call following the alarming call the mobbing call hereafter (**Figure 7**). During mobbing calls, gibbons occupied the crown canopy, which remained relatively concealed while calling and continued to survey the external environment.

The alarming call was composed of short simple syllables (aa notes). Different individuals in a group often repeated these simple syllables. The duration of an alarming call was 7–10 min. There was a difference in the duration of alarming calls among the three groups (ANOVA: $F = 8.42$, $df = 14$, $P < 0.05$), and the duration was longer in larger groups (A > B > C). Acoustic spectrum analysis indicated that the mean duration of a syllable in an alarming call was 0.078 ± 0.014 s and 520–1000 Hz. Acoustic frequency differed between groups, and a difference was detected in the highest and lowest alarming call frequencies between groups (ANOVA: $F1 = 17.34$, $df = 112$, $P1 < 0.01$; $F2 = 15.752$, $P2 = <0.01$).

The alarming behavior is the adaptation strategy of animals when they faced threaten from the habitat, especially for the primates and the group living animals. Alarm call of Hainan gibbon is simple, composed of only simple syllables and the frequency is low, but effective. This may be easier for young gibbons to learn, different gender and age stages of individuals have two sound signals, and indicate that the Hainan gibbon is a simple primitive species in the family Hylobatidae.

4. Discussion

Hainan gibbon is most rare and endangered wild animals that is only one in the world without zoos captive breeding has a very high protection value. In recent years, Hainan gibbon population is growing, but the suitable habitat area is too small, the low quality of food resources seriously limits the further recovery of its population. Zhang et al. [53] reported that the suitable habitat of Hainan gibbon decreased 35% in 17 years from 1991 to 2008. The protection regulators lack independence and administrative level is low in Bawangling nature reserve, the law enforcement power and management power are insufficient [54]. To make gibbon to get rid of extinction state, we must provide them with more habitat, together with the present situation of the vegetation and existing management measures in Bawangling nature reserve, we suggested that short-term is given priority to with local protection, lowland rain forest as the main measures to restore living habitat, medium and long term are focused on through the establishment of ecological corridors connecting the reserves in the central mountains, such as Yinggeling, Wuzhishan, and Diaoluoshan which have had the Hainan gibbon distribution, through the construction of the habitat corridors, connect the reserves and establish covers a variety environment characteristics of the protection of the reserve system. We can believe that under the joint efforts of all departments at all levels, the Hainan gibbon will have better development.

Current research results show that the population of Guizhou sub-nosed monkeys has no significant growth in recent decades and the active region is at elevations above 1400 m. So, the conservation of existing habitat is crucial, human disturbance from tourism should be strictly controlled. A total of 28 different plant species were eaten by the monkeys during different seasons in the whole year, some of the dominant species with high elevation distribution are important food resources for the monkeys, such as *Acer flabellatum, Pterostyrax psilophyllus, Sorbus xanthoneura* and *Magnolia sprengeri*. We should artificially plant important food resources for the monkeys to increase the food resources of the monkeys and expand the range of monkey activities.

5. Conclusions

Due to habitat fragmentation and human disturbance, the population and distribution range of Hainan gibbon and Guizhou sub-nosed monkey are drastically decreasing. The population of China's peculiar treasure endangered primates Hainan gibbon and Guizhou sub-nosed monkeys is extremely low. Distribution area was confined to an isolated island. It is very important to make strict protection measures for the endangered animals. To make the endangered species from extinction, the relevant research and conservation of protected species should be constantly strengthened.

Author details

Jiang Zhou[1*] and Huaiqing Deng[2]

1 School of Karst Science, Guizhou Normal University, China

2 School of Life Science, Guizhou Normal University, China

*Address all correspondence to: zhoujiang@ioz.ac.cn

References

[1] Zhou J, Wei FW, Li M, et al. Hainan black-crested gibbon is headed for extinction. International Journal of Primatology. 2005;**26**(2):453-465

[2] Liu ZH, Zhang YZ, Jiang HS, Southwick C. Population structure of Hylobates concolor in Bawanglin nature reserve, Hainan, China. American Journal of Primatology. 1989a;**19**:247-254

[3] IUCN. IUCN Red List of Threatened Species. Gland, Switzerland & Cambridge, UK: IUCN; 2010. Available from: http://www.iucnredlist.org. Accessed February 20, 2011

[4] Niu KF, Tan LC, Yang YQ. Altitudinal movements of Guizhou snub-nosed monkeys (*Rhinopithecus brelichi*) in Fanjingshan National Nature Reserve, China: Implications for conservation management of a flagship species. Folia Primatologica; International Journal of Primatology. 2010;**81**(4):233-244

[5] Xiang ZF, Wang L, Huo S, Cui LW, Xiao W, Quan RC, et al. Distribution, status and conservation strategies of the black-and-white snub-nosed monkey *Rhinopithecus bieti* in Tibet. Oryx. 2007;**41**:525-531

[6] Xiao W, Ding W, Cui LW, Zhou RL, Zhao QK. Habitat degradation of Rhinopithecus bieti in Yunnan, China. International Journal of Primatology. 2003;**24**:389-398

[7] Thomas O. Note on the gibbon of the island of Hainan (*Hylobates hainanus*, sp. n.). Annals and Magazine of Natural History. 1892;**9**(6):145-146

[8] Pocock RI. Observations upon a female specimen of the Hainan gibbon (*Hylobates hainanus*), now living in the society's gardens. Proceedings of the Zoological Society of London. 1905;**6**(2):169-180

[9] Delacour J. La systematique des Gibbons Indochinois. Mammalia. 1951;**15**:118-123

[10] DaoVanTien. On the north Indochinese gibbons (*Hylobates concolor*) (Primates, Hylobatidae) in north Vietuam. Journal of Human Evolution. 1983;**12**:367-372

[11] Groves CP. Systematics and phylogeny of gibbons. In: Rumbaugh DM, editor. Gibbon and Siamang. Vol. 1. Basel, Switzerland: Evolution, Ecology Behavior and Captive Maintenance. S. Karger; 1972. pp. 1-89

[12] Liu ZH, Yu SM, Yuan XC. The resource status of Hainan gibbon. Journal of Wild Animals. 1984;**6**(6):1-4

[13] Ma SL, Wang YX. Gibbon's classification and distribution in southern China - with the description of three new subspecies. Journal of Zoology. 1986;7(4):393-421

[14] Xu LH, Liu ZH, Yu SM. The Birds and Animals (Mammals) in Hainan Island. Beijing: Science Press; 1983

[15] Su B, Kressirer P, Monda K, et al. China's black gibbons' genetic diversity and molecular systematics study - noninvasive sampling DNA sequence analysis. Science in China (Series C). 1996;**26**(5):414-419

[16] Geissmann T. Taxonomy and evolution of gibbons. In: Soligo C, Anzenberger G, Martin RD, editors. Anthropology and Primatology into the Third Millennium: The Centenary Congress of the Zurch Anthropological Institute (Evolutionary Anthropology). Vol. 11. NewYork: Wiley-Liss; 2002a. pp. 28-31

[17] Brandon-Jones D, Eudey AA, Geissmann T, Groves CP, Melnick DJ, Morales JC, et al. Asian Primate

Classification. International Joumal of Primatology. 2004;**25**(1):97-164

[18] Deng HQ, Zhou J, Yang YW. Sound spectrum characteristics of Hainan gibbon (*Nomascus hainanus*) songs. International Journal of Primatology. 2014;**35**(2):547-556

[19] La Quang Trung, Trinh Dinh Hoang. Status review of the CaoVit black crested gibbon (Nomascus nasutus nasutus) in Vietuam. In: Nadler T, Streieher U, Long HT, editors. Conservation of Primates in Vietnam. Hanoi: Frankfurt Zoological Society; 2004. pp. 90-94

[20] Mootniek AR. Gibbon (Hylobatidae) species identification recommended for rescue and breeding centers. Primate Conservation. 2006;(21):103-138

[21] Roos C, Thanh VN, Walter L, Nadler T. Molecular systematizes of Indochinese Primates. Vietnamese Journal of Primatology. 2007;**1**(1):41-53

[22] Zhang YZ. Hainan gibbon (*Hylobates concolor hainanus*) is threatened. Asian Primates. 1992;**2**:6

[23] Deng HQ, Zhang MX, Zhou J. Recovery of the critically endangered Hainan gibbon *Nomascus hainanus*. Oryx. 2017;**51**(1):161-165

[24] Quan GQ, Xie JH. Notes on Rhinopithecus *Roxellanae brelichi* Thomas. Acta Theriologica Sinica. 1981;**1**(2):113-116

[25] Tan BJ. The status of primate in China. Primate Conservation. 1985;**18**(5):63-81

[26] Eudey AA. Action Plan for Asian Primate Conservation 1987. New York: IUCN/SSC Primate Specialist Group; 1987. p. 91

[27] Yang YQ, Lei XP, Yang CD, et al. Fanjingshan Research, Ecology of the Wild Guizhou Snub-Nosed Monkey. Guiyang: Guizhou Science Press; 2002. p. 96. In Chinese

[28] Bleisch WV. Primate research and conservation. In: Xia WP, Zhang YZ, editors. Mammalogical Society of China. Beijing: China Forestry Publishing House; 1995. p. 150

[29] Quan GQ, Xie JH. Research on the Golden Monkey. Shanghai: Shanghai Science and Technology Press; 2002

[30] Xiang ZF, Nie SG, Lei XP, Chang ZF, Wei FW, Li M. Current status and conservation of the gray snub-nosed monkey *Rhinopithecus brelichi* (Colobinae) in Guizhou, China. Biological Conservation. 2009;**142**(3):469-476

[31] Guo YQ, Zhou J, Song XH, Deng HQ, Qiu Y, Shi L, et al. The population of *Rhinopithecus brelichi* in Fanjingshan National Nature Reserve, Guizhou, China. Acta Theriologica Sinica. 2017;**37**(1):104-108

[32] Kolleck J, Yang MY, Zinner D, Roos C. Genetic diversity in endangered Guizhou snub-nosed monkeys (*Rhinopithecus brelichi*): Contrasting results from microsatellite and mitochondrial DNA data. *PLoS One*. 2013;**8**:e73647

[33] Yang L, Emily L. The particular foods resources and constituent characteristic of the vegetation in habitat of Guizhou golden monkey in the Fanjingshan mountain. Journal of Guizhou Normal University (Natural Sciences). 2002;**20**:18-24

[34] Zhou XM, Meng XH, Liu ZJ, Chang J, Wang BS, Li MZ, et al. Population genomics reveals low genetic diversity and adaptation to hypoxia in snub-nosed monkeys. Molecular Biology and Evolution. 2016;**33**:2670-2681

[35] Leighton D. Gibbons: Territoriality and monogamy. In: Smuts B et al., editor. Primate Societies. Chicago: The University of Chicago Press; 1987. pp. 135-145

[36] Carpenter CR. A field study in Siam of the behavior and social relations of the gibbon (Hylobates lar). Comparative Psychology Monographs. 1940;**16**(5):1-212

[37] Liu ZH, Jiang HS, Zhang YZ, et al. Field report on the Hainan Gibbon. Primate Conservation. 1987;**8**(1):49-50

[38] Jiang HS, Song X, Zhang J, et al. The Population Dynamic of Hylobates concolor Hainanus in Bawangling National Nature Reserve in Hainan Island. Unpublished report. 1999; (in Chinese)

[39] Chivers DJ. The Siamang in Malaya: A field study of a Primate in tropical rain forest. Contributions to Primatology. Basel: Karger. 1974;**4**:1-335

[40] Liu ZH, Zhang YZ, Jiang HS, Southwick C. Population structure of Hytobates Concotor in Bawangling Nature Reserve, Hainan, China. American Journal of Primatology. 1989b;**19**(4):247-254

[41] Liu XM, Liu ZH, Chen J, Jiang HS. The home range use and seasonal change of Hainan gbbon. Sun Yatsen University Forum. 1995;**14**(3):168-171

[42] Fellowes JR, Pui Lok CB, Zhou J, et al. Current status of the Hainan gibbon (*Nomascus hainanus*): Progress of population monitoring and other priority actions. Asian Primates Journal. 2008;**1**(1):4-11

[43] Zhou J, Li XC, Zhou ZL, et al. The application of GIS technology to the Hainan gibbons' conversation. Journal of Guizhou Normal University (Natural Sciences). 2009;**27**(4):22-29

[44] Chivers DJ. The swinging singing simians: Fighting for food and family in Far East Forests. In: Sodaro V, Sodaro C, editors. The Apes: Challenges for the 21st Century. Brookfield, Illinois, U.S.A.: Brookfield Zoo, Brookfield Zoo, Conference Proceedings, Chicago Zoological Society; 2001. pp. 1-27

[45] Guo YQ, Zhou J, Xie JH, Garber PA, Michael B, Ren BP, et al. Altitudinal ranging of the Guizhou golden monkey (*Rhinopithecus brelichi*): Patterns of habitat selection and habitat use. Global Ecology and Conservation. 2018;**2018**:e00473. DOI: 10.1016/j.gecco

[46] Deng HQ, Zhou J. Thirteen years observation on diet composition of Hainan gibbons (*Nomascus hainanus*). North-western Journal of Zoology. 2018;**14**(2):213-219

[47] Zhou J, Wei FW, Li M, et al. Reproductive characters and mating behaviour of wild *Nomascus hainanus*. International Journal of Primatology. 2008;**29**:1037-1046

[48] Geissmann T. Duet-splitting and the evolution of gibbon songs. Biological Reviews. 2002b;**77**:57-76

[49] Geissmann T. Gibbon songs and human music in an evolutionary perspective. In: Wallin NL, Merker BS, editors. The Origins of Music. Cambridge, Massachusetts: MIT Press; 2000. pp. 103-123

[50] Geissmann T. Evolution of Communication in Gibbons (Hylobatidae). Anthropological Institute, Philosoph. Faculty II, Zürich University; 1993

[51] Ruppell J. Vocal diversity and taxonomy of Nomascus in Central Vietnam and southern Laos. International Journal of Primatology. 2010;**31**(1):73-96

[52] Deng HQ, Gao K, Zhou J. Non-specific alarm calls trigger mobbing behavior in Hainan gibbons (*Nomascus hainanus*). Scientific Reports. 2016;**6**:34471. DOI: 10.1038/srep34471

[53] Zhang MX, Fellows JR, Jiang XL, Wang W, Chan BPL, Ren GP, et al. Degradation of tropical forest in Hainan, China, 1991-2008: Conservation implications for Hainan gibbon (*Nomascus hainanus*). Biological Conservation. 2010;**143**:1397-1404

[54] Xie Y, Yang SB, Wen YL, Su WB. Study on the status of conservation and management and suggestions on countermeasures for the Hainan Bawangling national nature reserve. Forestry Resources Management. 2009;**38**(3):22-26

Chapter 9

Modern State Law: Regulating Tradition or Protecting the Environment in the Mankon Kingdom of Northwest Cameroon?

Ngambouk Vitalis Pemunta and Ngwa Donald Anye

Abstract

Most African countries including Cameroon find themselves in a situation of legal pluralism and at crossroads with implications for the sustainable management of natural resources. Traditional institutions and knowledge systems have been hailed as invaluable mechanisms for the conservation of flora and fauna. This chapter examines the conflict between traditional institutions and State law in the hierarchically stratified Mankon Kingdom of the Grassfield region of Northwest Cameroon where the latter prohibits the harvesting of culturally valuable plant and animal species for myriad ritual ceremonies and for therapeutic purposes. It demonstrates that the lack of cultural sensitivity can be antithetical to conservation initiatives. In other words, there is the need to align current legislative regulations for the management of natural species with the traditional use of territory and gender roles as well as to raise the cultural and educational level of the population through sensitization on the need to conserve the natural environment on which their culture depends for its survival.

Keywords: legal pluralism, sociocultural factors, ritual, conservation, traditional society, environmental education

1. Introduction

Globally, the hunting of bushmeat[1] for food, ornamental, and medicinal products has put 301 terrestrial mammal species at the verge of extinction. The risk of extinction is higher in developing countries where rapid deforestation, the expansion of agriculture, human encroachment, and competition with livestock are accentuating the situation [1]. In Southeast Asia, 113 species of mammals are threatened by hunting. An estimated "13% of all threatened mammals are east of India and south of China, 91 in Africa (8%), 61 in the rest of Asia (7%), 38 in Latin America (3%) and 32 in Oceania (7%)" [1]. In Central and West Africa, the exploitation of bushmeat has reached alarming proportions. Although bushmeat is an all-time significant resource, in bushmeat-dependent communities,

[1] The term applies to all wildlife species including guinea fowl, monitor lizard, forest antelope (duiker), chimpanzee, gorilla, elephant, monkey, and other primates.

"traditional subsistence use of game is vulnerable to commercial and social changes" [1, 2]. The factors driving these transformations may include recent immigration, urbanization and the market economy's increasing demand and improvement in infrastructure allowing easier transportation of meat to markets far away [1, 2].

The Republic of Cameroon boasts of an estimated 9000 plant species. At least 156 of these plants are endemic. Of 409 species of mammals, 14 are endemic. The country also counts 2084 insects including more than 1500 butterflies. The country further has 542 species of fish, of which 96 are endemic. Other animal species include 330 reptiles and 200 amphibians [3]. Additionally, the country further has some 925 species of birds, of which 22 are endemic, while 249 plants are critically endangered [4]. The region where the Mankon Kingdom is found constitutes part of the World Wildlife Fund's (WWF) 200 worldwide ecoregions. The ecoregion of the Western Grassfields contains 35 restricted range endemic bird species, (third richest region for birds in mainland Africa). The region's habitat supports a number of endemic reptiles (10), endemic amphibians (55), endemic mammals (6), and an estimated 100 rare/endemic plant species [3, 5]. The Northwest Region is losing biodiversity at a rapid pace—37% of forest cover in 1958 shrank to 3.5% in 2000 [5]. Some species are generally thought to be at risk from the bushmeat trade or are currently rare or vulnerable or facing the threat of extinction. This means that they could easily succumb to the traditional pressures of hunting, thereby exacerbating their further decline and eventual extinction. In the Mankon Kingdom, the following wildlife species are at the verge of extinction: leopard (*Panthera pardus*), golden cat (*Profelis aurata*), forest elephant (*Loxodonta africana*), black colobus (*Cercopithecus preussi*), giant pangolin (*Manis gigantea*), and zebra duiker (*Cephalophus zebra*) [5]. The natural habitat of the Grassfield region is seriously threatened by a population increase of 3.4% per annum that has led to the creation of farms, unsustainable harvesting of barks of trees (for medicine), fires by grazers and farmers, unsustainable collection of wood for fuel, and construction as well as bushmeat hunting [3, 5, 6].

Despite the threat and likely extinction of certain wildlife species, Cameroon, however, remains the second richest country in Africa in terms of the diversity of its primates [7]. To strike a balance between conservation and development, the government of Cameroon enacted a forestry policy and its decree of implementation in 1994 and 1995, respectively. The country further adopted a policy relating to environmental management as well as an accompanying legal framework in 1996—Decree on Environmental Impact Assessment [2, 5]. The effective implementation of these policies by Cameroon's Ministry of Forestry and Wildlife and of the Environment and Nature Protection is, however, dogged by inadequate diffusion among local frontline staff at the grassroots and ill-informed community members who unknowingly harvest protected plant and animal species as status symbols and for other sociocultural purposes. The exploitation of rare plant and animal species shows the chasm between the adoption and the implementation of culture-insensitive laws for biodiversity conservation. For example, by outlawing the hunting of certain species of animals, Cameroon's 1994 Forestry Law has created a leeway for community members to be arrested by the Ministry of Environment and Nature Protection during the celebration of culturally significant events when products from protected wildlife species are on display.

Mankon is one of several Western Grassfield chiefdoms that shares most of its basic ideological and social repertoire with other chiefdoms of the region. This includes sacred kingships, the belief in misfortune and pollution, and the special status of certain animals like the leopards, among others [8]. We were surprised during previous research visits to the Western Grassfield region when we observed that interviewees in authority including the Mankon Fon (King), notables, clan heads, quarter heads, and other influential persons were exhibiting either some animal skin or some of their parts. The Fon, for instance, does not generally sit anywhere without animal signs on his outfit or animal skin used as foot mat or chair cover. The increasing scarcity of most of these animal species that are used as status and ornamental symbols and Cameroon's conservation laws addressing the protection of certain threatened culturally significant species will certainly put the people in a double-bind situation. Why have people despite the ban by the Cameroon government continued to display certain in-/tangible wildlife species? This ban has put the people of Mankon and other traditional societies in a dilemma when it comes to the use of parts of plants and animals in their environment. It also raises questions about the effectiveness of traditional mechanisms of conservation/management of natural resources. Natural resource "management" has to do with how species are controlled or directed as a resource, while "use" refers to the functions these species serve or the uses to which they are put. The distinction between "use" and "management" of wild animal species is critical for assessing their integration into community forestry activities. For community forestry to incorporate wild animals, provision must be made for their effective management, not merely their use. In a context where State laws supersede local cultural norms and practices, what is the relationship between gendered wildlife use and the socioeconomic, political, and religious life of the Mankon community?

This chapter examines the conflict between State law prohibiting the use of certain culturally significant flora and fauna species as status symbols and for sociocultural purposes in the hierarchical Kingdom of Mankon in the Western Grassfield region of Cameroon. Unsustainable exploitation has virtually eliminated wildlife, and vulnerable species are threatened with extinction. Concurrently, most of the symbolic values associated with the use of royal wildlife are persisting. The chapter demonstrates that the lack of a gender lens and cultural awareness in the formulation of conservation legislation can be antithetical to the protection of the environment and the management of natural resources. The chapter argues that tightening the rule in the name of protecting the use of wildlife species can entail severe restrictions in traditional rights. The chapter calls for a synergy between current conservation regulations for the management of flora and fauna with traditional gendered use of territory, especially the use of certain symbolic animal species. It suggests that raising the cultural and educational levels of the population could lead to behavior change, but there is also the need to adapt legislation to local gender roles and cultural norms to ensure the protection of the environment and sustainable development.

Educational attainment is considered a quintessential tool for environmental protection as well as a determinant of environmental preference. We argue that an individual's level of education strengthens his ability to "receive, decode and to understand information processing and interpretation have an impact on learning and behavior change towards the natural environment" [8, 9]. According to the Oxford English Dictionary, education consists of the "systematic training and instruction designed to impart knowledge and develop skill" [10]. In every society, it is the catalyst for economic and social changes [11] as well as for changes in attitudes and behavior. It goes beyond the simple "acquisition of knowledge and includes the ability to evaluate that knowledge" [9, 12]. As succinctly put by the World Conservation Union, environmental education encompasses changes in behavior and the idea that knowledge will induce personal, societal, and global changes [13]. Learners will be afforded an "opportunity to gain an awareness or sensitivity to the environment, knowledge and experience of the problems surrounding the environment, to acquire a set of values and positive attitudes, to obtain the skills required to identify and solve environmental problems and, the motivation and ability to participate" [9, 12].

2. Method of study

This chapter examines the conflict between modern State law and the cultural use of prohibited wildlife species in the cultural universe of the Mankon people of Northwest Cameroon. Multilocal ethnographic fieldwork [14, 15] took place in all clans and neighboring villages of Mankon in the Northwest Region of Cameroon. It lasted for 6 months (January to June 2010). We used semi-structured and open-ended interviews as well as observations. Qualitative methods were deemed suitable because of their ability to explore the meaning and understanding that people attach to phenomenon [16]. Firstly, they provide a "thick description" or depth of understanding to complement the breadth of understanding afforded by quantitative methods. Secondly, "they elicit the perspective of those being studied, explore issues that have not been well-studied, test hypotheses, or evaluate the process of a phenomenon" [1, 17].

"Open-ended" questions enabled us to explore different perspectives and methods to generate breadth of knowledge and depth of experience in order to understand and appreciate the phenomenon under study [17]. Lastly, it provides excellent understanding into people's opinions, experiences, and perceptions from their own point of view—including subjective experience [18–23] such as perceptions of forest and wildlife.

The exploratory phase combined qualitative and quantitative ethnographic methods aimed at describing the cultural model of wildlife use. We elicited the cultural significance of wildlife use as a determinant of some cultural values in the community. During the initial phase of fieldwork, passive observation with minimum interaction between us and the object of study was helpful. As passive observers, we established residence in Mankon, made new acquaintances, took part in community activities, worked to establish rapport, and attempted to understand how considerations of wildlife use play into everyday life. Our main task was to make a systematic record of day-to-day interactions, observations, and informal preservation by writing field notes on a daily basis ([23], pp. 180–207). As part of an iterative process, writing field notes was helpful to us because we could identify important questions and domains of life that needed to be explored in detail. Passive observation was an appropriate method for addressing sensitive questions such as hunting following the ban on certain animal species. It extended the internal and external validity of the study by helping us understand the meaning of observations, and it helped us to formulate sensible questions for later stages of the research ([23], p. 141).

We obtained research authorization from the Head of the Department for Anthropology in the University of Yaounde 1, Cameroon. Participants were debriefed. They signed a consent form but were also free on when and how to withdraw from the study if a need arises.

In the next sections, we first present the dynamic socioeconomic and political context in which State law takes precedence over customary land ownership, including ownership of the forest and wildlife species inhabiting it. The second part situates the research area and the local sociopolitical organization and examines traditional mechanisms of conservation in Mankon. The third part examines cultural representations of wildlife, especially royal wildlife in the sociocultural universe of the people. The last part is the conclusion.

2.1 Legal pluralism and the management of natural resources

This section demonstrates that in Cameroon, the coexistence of State law and customary law regulating land and the resources on them pits the government against local communities and tends to undermine community rights over their natural resources.

Cameroon stands at a crossroad between modern State laws and traditional authority. This legal dualism has led to a conflict of laws, especially when it comes to the conservation of natural resources. In most parts of Cameroon, the enduring power of traditional authorities still holds sway. Consequently, the word of the local Fon, chief, or sultan/lamido still holds more weight than that of the government. In all the countries of the Congo Basin, wildlife remains important to all the chiefs. Local communities and the state, however, view wildlife differently. At the local level, it is used for food and for medicinal and cultural purposes (especially in rituals and as emblems by traditional dignitaries), and it is traded through barter or commerce [24]. Through various ritual actions executed by elders, they transmit this knowledge from one generation to the other. Apart from changing environmental conditions, State laws that prohibit the killing of particular species of animals may transform these ritual actions. The legal ban on certain species of animals suggests the loss of cultural traits.

The State of Cameroon claims monopoly over all land and the natural resources therein, while traditional authorities like the Fon of Mankon are custodians of land and natural resources at the local level for their people. Access to land, usually through lineages, is a "communal" right, while colonial and presently postcolonial rules are experienced through the network of power relations known as the "customary" [25]. Mahmood Mamdani has baptized this system of indirect rule that uses chiefs/Fons as auxiliaries of the administration (citizens and subjects) as institutionalized despotism [25]. This implies that although bestowed with traditional authority, the Fon remains subservient and an auxiliary to the modern administration.

Prior to the Franco-British partition of Cameroon and subsequently the colonial encounter that tremendously transformed property rights, traditional institutions headed by traditional rulers and members of their councils managed natural resources including the forest. This was done in tandem with customary norms, practices, and beliefs [26, 27]. This resonates with calls for the integration of traditional ecological knowledge into modern natural resource and environmental management systems. Serra Jeanette Hoagland [28] has, for instance, pointed to the need for a dualistic perspective where traditional ecological knowledge and western science are integrated into natural resource management. This implies the integration of different ways of knowing (different knowledge systems) into natural resource management decision-making and strategies. Like in other kingdoms in Africa (Ghana, Nigeria, Tanzania, and Zimbabwe), uncultivated land was communally owned [24, 26, 27]. There was minimal forest resource exploitation, and it was mainly done for subsistence purposes. Traditionally, the forest served as a source of food, medicine, fuel wood, water, construction materials, and place of worship (spiritual sanctuary) [29]. The traditional ruler granted permission for hunting after determining where and when the hunting expedition will take place as well as the type of animals to be hunted. Those who entered the forest without his permission could be severely punished and, in extreme cases, excommunicated from the community [30, 31]. Low population pressure made the impact on the environment to be low [32]. Today, resources such as wildlife have progressively been subjected to government control at the expense of traditional landholders/local communities [30–32].

The overlapping power structure between the modern state and traditional authorities has led to a conflict of laws. When it comes to the management of the environment and natural resources, these mutually contested rights that characterize land tenure in present-day Cameroon have been dubbed by Phil René Oyono as a "deep *conflit de langage* (in French) [32]. According to him "the *conflit de langage* (conflict of language or of discourse)" is specifically "between the state and local communities on land and forests ownership and on the regulation of access to natural resources" ([32], p. 115). This implies that there are conflicting discourses as well as claims and counterclaims between local authorities and the state about who really owns the land and the forest [29, 32, 33]. Following the complete confiscation of their resources by the state, members of local forest-dwelling communities are increasingly feeling a sense of alienation and deprivation [34–36]. This implies that Cameroon's 1994 Forestry Law is not culturally sensitive. It is therefore a legislation against local people because it denies them access to their natural resources [36].

The 1994 Forestry Law does not identify the legal status of customary land interests. The Cameroon government has failed to acknowledge customary landholding. This is evidenced by the difficulty that customary landowners face in formally registering and securing their landholdings [27, 33]. This scenario has rendered rural Cameroonians deeply insecure in their land tenure. They are squatters on their own land. Land ownership continues to be reckoned in terms of "development" [27, 29, 30]. Cameroon's land legislation has, therefore, fallen short in social and developmental respects concerning its treatment of customary tenure. By failing to recognize customary land rights, the provisions of the 1994 Forest, Wildlife and Fisheries Regulations in Cameroon "have added significantly to the wrongful demise of customary land rights through unnecessary and patently rent-seeking ways" as instantiated by its designation "of the most valuable forest resources of its citizens as its own private property. The outcome also shows that sustainable forest conservation and management of use are also being affected" [4, 36]. This highlights the enduring role of local conservation practices.

2.2 Traditional mechanisms of conservation

The Mankon people of Cameroon's Western Grassfield region find themselves in a new economic and political dispensation. There is the intrusion of State laws regulating the hunting of particular large mammals including the lion and elephant that are symbolic of royalty. In this new context, State authority has taken precedence over indigenous land rights and the natural resources therein. This State authority is instantiated and concretized through the declaration and implementation of an environmental policy (The 1994 Forestry Law) that defines exclusionary rights [37]. It has imposed limitations on animal species that can be legally hunted as well as the tools that can be used during hunting (**Table 1**).

There is power inequality between the state and conservation organizations on the one hand and, on the other, between them and the Mankon community. This power inequality is about who exercises power, when, and how. This is evident in the competing knowledge systems at play in the framing of sustainability between the indigenous Mankon people who have for decades been stewards of their environment and the government and conservation organizations over the environment [37]. Against this backdrop, the role of indigenous peoples and of local cultural norms in conservation has been questioned [37–40]. Practices including resource rotation, food taboos, and restrictions on harvest limits are allegedly the outcome of optimal harvesting and not conscious conservation efforts. Scholars have taken

Animal category	Definition
Class A	Totally protected No hunting No trade or use as pets Authorization from Wildlife Service of MINFOF required for their capture
Class B	Protected Hunting allowed with hunting permit
Class C	Partially protected Wildlife Service of MINFOF regulates their exploitation
CI	Trade restricted in tandem with the Convention on International Trade in Endangered Species (CITES) Examples—elephant tusk, montium chameleon, African rock python, and *Prunus africana* (pygeum)

Compiled by author with information from Verina Ingram and Nsom A. Jam [5].

Table 1.
Categories of protected species according to Cameroon Law No. 1996 relating to environmental management.

issue against the conflation of sustainability which results from small population sizes and therefore minimal pressure on natural resources with conservation, that is, the intentional restrictions of short-term goals for the achievement of better long-term gain [38]. The crux of this debate is about the role of indigenous peoples in conservation initiatives [38–42].

Despite the assumption that rural people traditionally harvest resources sustainably [41], the sustainability of subsistence economies, requiring internal controls on population and exploitation, has not been demonstrated to exist [40, 43]. Traditional subsistent participants who have entered commercial markets dominate much of the current bushmeat trade. One of the primary concerns of those working on the bushmeat issue is the very real threat that this poses to communities that are largely dependent upon wildlife for meeting their primary protein needs. Truly sustainable subsistence hunting communities like Mankon are difficult to identify, but recent research has shown that it may be feasible to maintain sustainable levels of exploitation for subsistence as long as significant controls on hunting access and methods exist along with controls on trade and demographics (i.e., immigration to local community) [42]. These may include prohibitions on the slaughtering of the leopard, lion (/sam-bang/), cheetah, and others. Do indigenous people (including the Mankon people) have mechanisms for conserving their environment? We set the context of study before examining traditional mechanisms of conservation and the Mankon people's myriad perceptions of the forest.

2.3 The study site

In this section, we describe the local sociopolitical organization and highlight the sociocultural significance of wildlife parts to the Mankon cultural universe. Located in the Northwest Region of Cameroon, Mankon is part of the Grassfield chiefdoms where the threat to tropical animal species has been reported [6]. It is found on an altitude of about 1000 m, above sea level. It shares boundaries with tribes including Bafut in the North. The Meta, Ngyembu, and Bali Nyonga Fondoms are situated in the east, and Nsongwa, Mbatu, and Akum are in the south.

2.4 Mankon in the Western Grassfield region: Cameroon and Africa

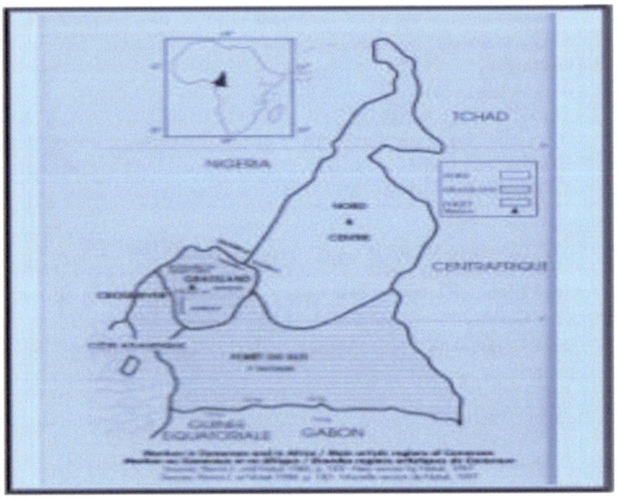

Source: Notue [8, 25].

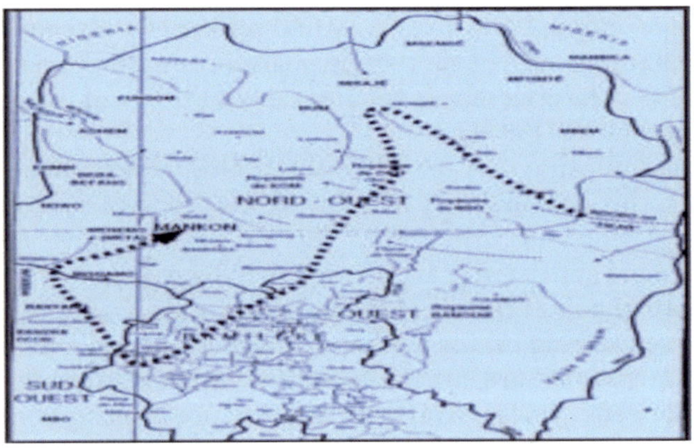

Source: Notue [8, 25].

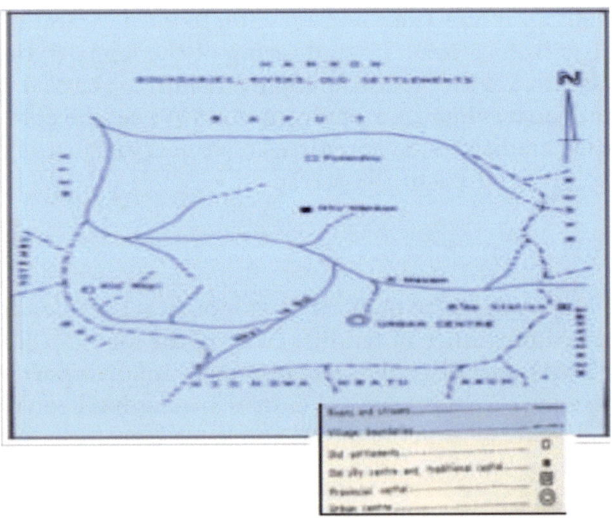

Source: Eballa et al. ([44], 1984:1).

Mankon is bounded in the east by Mushugu and the villages of Mandankwe and Nkwen: in the North by Bafut; in the west by Meta, Ngyen-mbo, and Bali; and in the south by the villages of Mbatu and Nsongwa. Mankon has an estimated land area of about 315 square kilometers and an estimated population of about 50,000 inhabitants (estimate made in 1984). On the average, therefore, the population density is about 158 persons per square km [45].

Despite the impact of multiple raids in the nineteenth century, Mankon retained its cohesion. It also preserved its social organization that is based on nine clans. After the Ba'ni raids, the people settled in a dense defensive settlement on the north bank of the river Mezam. It is reputed for having retained its basic clan structure [45]. It was originally an amalgamation of three villages (Mbatu, Nsongwa, and Chomba), and it is one of the principal ethnic groups that constitute the larger Tikar group of people.

Mankon has a demarcated territory, a population of diverse origins, well-defined traditional institutions for the regulation of conflicts, and a military force under the leadership of a sacred sovereign personality, the Fon. The *ntoh* (capital) is the royal palace and is the main cultural, administrative, and religious center of the community [6, 45]. Aspects of the belief system of the people are perceptible in the cultural events they perform before, during, and after important traditional ceremonies. Prominent among these ceremonies is the annual cultural dance event (*Abüng a Fo*), which is a replica of its political, economic, and social structures.

Because of its hilly topography, the climate is generally cold and windy. The alternating dry and rainy seasons of the year give room for subsistent farming leading to crop yields, which are mainly sold on traditional market days and Saturdays. These traditional market days attract people from other neighboring villages who converge to buy or sell food items.

Mankon is surrounded by lowland forest on the western and southern sides. The savannah, shrubs, and other tree types grow on the hilltops and slopes, while the elephant stalk (*nkünka/atasong*) and watershed trees are visible in the valleys and the lowlands. Rain forest trees can be found in the hinterlands. This is evidence of the fact that high equatorial forest trees once grew there. The ecosystem has reportedly witnessed depletion. This is the result of human activities like regular burning to create farms and exploitation by hunters, herdsmen, and farmers.

As a measure to reverse the threat of watersheds drying up, the Ministry of Wildlife and Nature Protection has embarked on numerous reforestation projects. There is, however, a concern for the tree types planted for this purpose. Most people grow *Eucalyptus* trees for various needs like fuel wood, roofing, poles, and other uses. Contrary to the importance of *Eucalyptus* trees, it is more sustainably useful in wetland areas since it is known for its huge water nourishments for growth. Therefore, a semiarid region like the Northwest Region will probably need other friendly tree species to rejuvenate its ecological system as already noted by officials of the ministry concerned. A further shortcoming with these projects is the conflation of community ownership with sustainability. As Victor Agah-ah Mah states, "The implementation of these projects disregarded the traditional beliefs and practices of end-users and engendered loss of access to shrines, groves and forest-based or water-based resources without providing alternatives. ...Increasing temperature and reducing amount of rainfall result in a greater incidence of bushfires, which threaten the sustainability of some community-managed projects" ([46], p. 5).

2.5 Traditional social organization

The Fon of Mankon is also the clan head of the *ntoh* clan—the royal clan of the village. The Fon presides over rituals, the war council, and the council of notables whose members are chosen according to lineage. The Fon has lots of power and is an authority above all other governing institutions in the land. He mediates in all litigations but for those involving murder and treason [6, 45]. The former cases fall under the jurisdiction of administrative courts. Apart from that, the Fon also heads all traditional societies that rule community life.

Source: Anye [6].

He is also part of an authority known as *Kwifo*, which represents the whole land in spiritual matters. Members of *Kwifo* decide and the Fon executes. In case the Fon passes on, they enthrone a new one. The *Kwifo* is endowed with lots of spiritual powers. They perform most of the sacrifices and are believed to be in direct communication with the ancestors of the land.

Christianity, Islam, and indigenous religion are practiced in Mankon. Most of the people are syncretic Muslims and Christians. Like most African peoples, they hold that the gods of the land exist everywhere and that the achievements of people are based on the protection they get from their gods and ancestors. Hence, in order to maintain a cordial relationship with these supernatural forces, they offer sacrifices and pour libations amid prayers or incantations [47]. The traditional religion still has an important role in the protection of their activities, environment, and themselves. Sacred sites as well as ritualistic objects accompany ceremonies and celebrations ([6, 8], p. 36).

2.6 Traditional mechanisms of conservation and perceptions of the forest

Traditional authorities in the Western Grassfield communities in Cameroon often rewarded people who captured certain animal species with traditional titles. The hunting down of wildlife species and the use of most wildlife parts are contrary to modern norms of conservation. It highlights the need to protect species threatened with extinction. If the species concerned are not protected, even the culture of the people is under threat once these species have become extinct. Other consequences include the progressive alteration of the traditional cultural pattern resulting in changes in habits and capabilities acquired by people as members of the society or lead to widespread environmental damage. People need education on the unfettered consequences of environmental degradation and the ruthless hunting down of wild animals.

Traditional institutions, knowledge, and practices are believed to have positively impacted on the sustainable management of the sacred forest in Mankon. The *Fo'* (King) and the secret society, *Kwifo*, and the Mankon traditional council are the major actors protecting the environment of Mankon in general and the sacred forest in particular. Traditional institutions, knowledge, and practices as well as religious values have been recognized to influence people's behavior and to significantly contribute in the conservation of forest [47].

2.7 Totemism

In the past, totemism served as one of the major traditional conservation tools for conserving myriad plant and animal species. Similarly, cultural beliefs in totemism, taboos, and sacred forest around villages adjacent to the Korup National Park in Cameroon have, for instance, been hailed for protecting the national park. Local bylaws that are binding on local inhabitants inordinately contribute to the rational use of natural resources [40]. This suggests that community norms remain sacrosanct to conservation initiatives. To ensure the sustainable use of both flora and fauna, there are taboos in place against the use of certain animal and plant species. Although material and symbolic approaches have widely divergent interpretations of people's behavior toward animals, both schools agree that individuals are often not conscious of the "true" reason behind their conduct. Indeed, as Mary Douglas [49] notes, followers of a cultural tradition often seize on secondary rationalizations, such as health concerns, to explain their tabooed beliefs and behaviors. Her distinction between the sacred and the profane/clean and unclean and the overall notion of contamination that are used to maintain the social order and to shape human behavior [49] can be useful for understanding taboos and their power in conserving the environment in local communities.

Project planners and managers need to examine carefully all taboos and related beliefs and to avoid making assumptions that ignore cultural variation within an area. To assume that one group's taboo on a particular type of fauna is shared by neighboring peoples, or even among members of the same family, may cause community forestry planners to shun species of wild animals that might otherwise be excellent candidates for development projects. This point is particularly relevant when outside influences have combined to make previously forbidden foods acceptable, as with deer in the tropics [50] or when previously acceptable foods have been discarded, as with mice and rats among the Maraca of Colombia. In addition, some wildlife species were adopted in the Mankon community and treated like family members. They often lived and, at times, ate with humans. For instance, dogs were often accorded burial and mourning rites. In many Indian tribes, people might be known as the father, mother, brother, or sister of such and such a dog.

2.8 The sacred forest

A sacred forest is a forest reserved by traditional authorities. Sacred forests are governed by unwritten by unwritten cultural norms and are used as sites for the performance of communal rites meant to honor the ancestors. They are believed to serve as a protective shield against any harm or calamity [47]. They contain varied plant and animal species. Most sacred forests are located around the palace and within protected areas. Community efforts at natural resource management in Mankon started with the creation of a forest close to the royal palace by the Fon in the early 1950s. This forest covers about five (5) hectares of land. For sacrifices, the Fon also uses another sacred forest called kekfure. It is also used for the Fon's recreation when he concentrates and communes with the ancestors. This forest is reputed for unique wildlife species not found elsewhere in Africa, notably some rare species of birds. Sacred forests are areas of valuable importance to the community in terms of ancestral burial grounds, valuable tree and animal species, etc. It is highly protected in the belief that immortal and omnipresent forces surround the forest area. As stated by various village elders, in the past this forest served as a security cordon against a possible attack by surrounding villages or invaders. People are prohibited from exploiting flora and fauna from this forest. It is believed that an individual could be affected by the anger of the spirits of the ancestors in case of violation of the rules and regulations [47–50] governing this forest.

2.9 Harvesting of medicinal plants

Cultural norms also surrounded the harvesting of medicinal plants by both traditional healers and non-healers alike. The parts of herbs required for therapy such as barks, leaves, roots, and quantity had to be harvested at particular periods of the day, usually in the morning/evening or on particular days of the week [50–52]. The aim was to ensure their rapid regeneration. Cultural prohibitions around harvesting are advertent measures of conservation.

> *Once the medicine had been found, rituals for harvesting—such as prayer and thanks to the ancestors—were performed. In most cases, the person would only harvest the part of the tree or plant (such as bark, leaves, roots etc.) that he or she needed. The rest of the tree would be saved for future use. Medicines were harvested at particular times during the year, mostly during the time when harvesting would do less damage to the tree or plant. The 'just in time' nature of use meant that people only harvested the medicines they needed* [32, 52].

2.10 Symbols of royalty

In the Mankon Kingdom and most Western Grassfield kingdoms, certain animals—leopards, buffaloes, tiger, and elephant—are considered as symbols of royalty. Accordingly, whoever captures or kills one of these is decorated and elevated in status. According to Lewellen, [...] symbols must serve a dual purpose: they must be at once particularistic, serving to unite the group and maintaining its unique identity, and universalistic, legitimizing it as an agency of power to the great majority of outsiders [53]. According to David I. Kertzer, true symbols have three properties, "[...] condensations of meaning, symbols are multivocal, finally, true symbols possess ambiguity, so that they can never be fully defined; they have no precised meaning" [9, 53, 54].

2.11 The leopard (*Panthera pardus*)

It is a large elegant and powerful feline, with elongated forms. Because of its cunning attitude, power, and ferocity, it is one of the most feared animals and, as such, one of the most respected in the African jungle [8]. The Mankon attributes to this creature many powers. It is associated with political and judicial authority and perceived as symbolizing strength, the power of the Fon, prestige, and the greatness of royalty. The Fon is indeed called "leopard" and his children "those" of this animal [8, 67].

Source: The authors.

In representing the feline as portrayed on works of art, the artist is said to bring out (using formal means) the qualities and virtues of the wild animal (strength, ferocity, and rapidity, among others). He seeks to simplify, stylize, and emphasize what he considers meaningful: emphasis is on the simple thickness of its body and head. The decorative design or pattern of the leopard are found on various articles and used in the celebration of enthronement and succession rites or during ancestral and fertility cults, for the prestige of the Fon: in a nutshell, for everything related to maintaining political authority and social cohesion [52]. Not only does the panther appear in the form of figurative motifs, but its hide and teeth are used for symbolic objects or ornamentation.

2.12 The elephant

It is also associated with power, command, and plenty [8]. The elephant like the leopard is said to be a symbol of royalty. The body parts of these animals are used in many rituals and cults of powerful secret societies within the Fondom of Mankon. The beast appears on various objects used by members of this society—receptacles, pipes, drums, masks, and seats, among others. When this animal is represented on masks and pipes, it incarnates the indomitable force of nature that can be harnessed and used for various profitable purposes by an individual or the community [8]. Moreover, in Mankon as elsewhere in the Western Grassfields, men (and especially the Fon) are said to form alliances with animals in order to have a double, triple, or multiple existence. This is in addition to having all the qualities, strengths, as well as weaknesses of the chosen animals to act in life efficiently. The elephant is not only a source of inspiration, as a plastic theme for Mankon and Grassland artists, but also a source of ivory and hair, which are used to make various works of art. Ivory is of great economic value and a sign of wealth. The marketing of this strategic product is perceived as the monopoly of the Fon of Mankon. Ivory products, luxury articles, or articles for prestige have for long remained the prerogative of the Fon and important personalities of the Fondom. Apart from the elephant, the buffalo is said to evoke strength, courage, and vitality. Like the elephant, it is a royal animal. This explains why the Fon has the right over a killed buffalo. He rewards the hunter who brings it back home and to the palace as booty.

2.13 Symbolism and the multiple uses of wildlife in Mankon

Wildlife is harvested daily across Africa using legal and illegal means as well as sustainable and unsustainable ones. In this section, we examine the views and opinions of respondents on how they treat wildlife—wild animals and royal animals alike. This includes the patronage of royal animals, the uses of royal wildlife, the royal animal and environmental sustainability, the use and trade of wild animals in traditional medicine, and other purposes.

2.14 Perceptions of wildlife

According to wildlife officials in Cameroon's Ministry of Forestry and Nature Protection, most animals in Mankon, including the African forest elephant (*Loxodonta cyclotis*) and the chimpanzee (*Pan troglodytes*), are endangered species. According to respondents, the community loves wildlife because it is a good source of protein; wildlife activity is useful for the ecological chain. In the past, it was used to symbolically determine the prowess of the traditional military force of Mankon. They were consequently treated with great respect.

In the course of migrating to settle in their present site more than a century ago, successive leaders distinguished themselves through bravery on the battlefield and hunting ground. Hunting demonstrated to an extent their resistance at the war front. In dynastic Mankon, wildlife hunting was a great social phenomenon. Today many human activities have encroached on the natural habitat of wildlife. Most trees have been cut down, and sometimes there is scarcity of food and water. Because of this human habitat encroachment, for instance, *Ngwabah* (an animal said to be in the family of tigers), that existed for a while in the neighborhood of Mbinfibieh in Mankon in the 1950s and caused aggression and terror to the community is now only in historical memory:

... when we cite the example of Sacred Heart College Mankon [...] the college area use to be the hunting ground of the Fon of Mankon [...] in those days when I was growing up as a young man, we had giraffes, antelopes, deer, and other wild animals that existed in the Mankon village [...] when you moved along the road you could see some of them inside that bush [...] but from the time they used that place to build the school, with electricity in the whole place [...] you cannot even see a rat mol there; so the children growing in Mankon today will never know that before this time giraffes, antelopes, dares and all the other types of wild animals existed in Mankon [...] so there is need for preservation to let the upcoming generation see and know the different types of animals.... (Aborengong John)

The Fon of Mankon is believed to be a patron of all wildlife within his area of jurisdiction. This reflects the royal highness' personal interests. While sometimes represented clearly as some human ruler, at other times, he is believed to appear more like a leopard, which is valued as very brave, ferocious, aggressive, intrepid, great watcher, courageous, and active. The most common representation is the one in which the Fon is identified in his palace same as in public places with certain wildlife parts like the horns of an elephant, its teeth, the skin of a leopard, and other parts like the feather and the hair of other royal wildlife. Other members of the Fon's entourage also have wildlife with which are identified. For instance, notables drink from buffalo horn cups. The designs of fabrics worn by notables carry the image of leopards to highlight authority and belonging to the class of power. The animal skin shown in the photo illustrated above depicts the use of some wildlife products. Traditional regalia and bags like the ones depicted below are valued in traditional ceremonial settings (**Figures 1–4**).

The bags depicted in the pictures above showcase wildlife products used not only as cultural items but also for decoration and ornamentation.

Any wildlife captured by Mankon citizens is taken to the palace as a gift to the highest authority of the land. Apart from the possession of some wildlife by the Fon, other species harvested as wild meat are for subsistence and noncommercial purposes. Some of these are animals under the category of non-protected species. They are hunted with the use of traditional snares, nets, registered guns, traps, etc. and for this purpose. However, individuals who have the right of access to the forest also hunt wildlife, including protected species for subsistence [43].

Estimated values for the wild meat trade in the formal literature are limited, perhaps due to the small number of harvest and trade activities in Africa that are considered legitimate and for subsistence. The importance of wildlife as a source of protein, religious significance, cultural value, medicinal use, and income to rural African communities remains true in the present day as it was during pre-colonial times [55, 56]. State regulation and the undermining of traditional user rights have however modified traditional values [55]. This has resulted in the perception of wildlife as an open-access resource and, as expected, overexploitation [56]. Where large-scale commercial markets have developed, the focus of hunting activities has

Figure 1.
Fabric in images featuring wildlife inscriptions. Source: The authors.

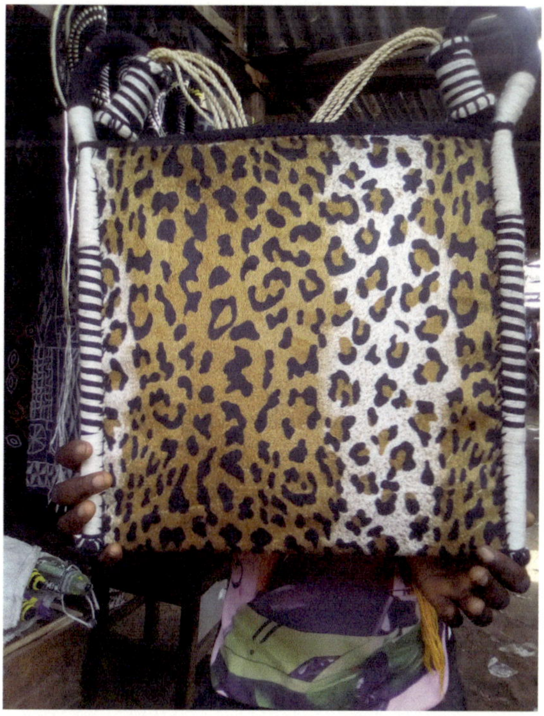

Figure 2.
Fabric in images featuring wildlife inscriptions. Source: The authors.

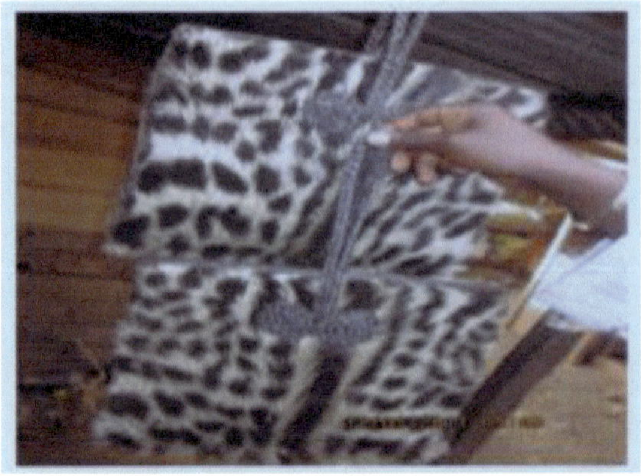

Figure 3.
Bags made out of wild cat skin (African civet). Source: The authors.

Figure 4.
Bags made out of wild cat skin (African civet). Source: The authors.

tended to concentrate on specific species [55]. In some communities, traditional law dictates that wildlife once killed becomes the property of the hunter or the proprietor of the hunting implement (gun, snare, and crossbow) [56]. This provides the necessary motivation for killing as many animals as possible to secure property in unregulated systems [57]. Several projects in East and Southern Africa have sought to mitigate this type of problem by granting specific rights of wildlife use to local communities [58]. The overall impact shows that in Southern Africa community wildlife use "has had significant impacts on the development of natural resource management regimes at national levels and added significantly to the sustainable use debate at the international level" [26, 59]. The impacts go beyond "conservation issues into the realm of human rights, democracy and constitutional reform" [59]. As stated by one community member:

> *We are now citizens who own or at least have control over our land and the benefits that come from using it. Government is becoming a partner in our development and officials in our area are starting to feel that they should work for our benefit. The company that hunts in our area is also becoming our partner. And all this means that in the eyes of our neighbors in Zimbabwe and Zambia we are no*

> *longer refugees, poor cousins whose land is no-mans land where anyone can do as they want. We are also now people in our own right* ([59], p. 26).

Respect for community norms and values will lead to the transformation of villagers from poachers to protectors of wildlife and plant species as a viable solution to participatory environmental conservation that give user rights and take the concerns of local forest-dwelling communities seriously [35, 59]. Some communities have, however, perceived the benefits from wildlife meat and products to be minimal. This demonstrates that there are always winners and losers in conservation initiatives [35].

2.15 Royalty as custodian of wildlife

Over the years, the royal family embraced a couple of wildlife as favored symbols of authority. Formal portraits from the eighteenth century onward have reportedly shown successive Fons and other notables posing or appearing with the aforementioned royal animal parts. From Fon Angwafo II (1866–1919) till date, most of the royal wildlife merit their own portraits as their parts are either portrayed and drawn or used almost permanently in the palace and public places wherever the authority of the land is represented. For instance,

> *...The animal skin which the predecessor of the Fon used, the Fon still has to use because if a citizen of Mankon catches a leopard, tiger, buffalo and other wild animals, they carry it to the palace and when it is slaughtered they preserve its skin...and they remain there forever and can be destroyed only by fire and other similar disasters.... (Pa Tumanjong Paul)*

During occasions where traditional dances are exhibited in Mankon, most of the traditional attires worn by either the dancers or other participants are adorned with the portraits of wild animals and other wildlife parts. These animal parts include the tail, horn, and skin. Similarly, during communal ritual performances including the annual traditional dance (*Abüng a Fo*), the Fon wears a cap made of elephant (*/usen/*) hair and tail. This cap was reportedly inherited from his predecessors who obtained it upon their successive accession to the throne. It is believed that in the past, all captured royal animals had their parts preserved by royalty for posterity and as symbols of authority. As one of our key respondents pointed out:

> *...The animal skin which the predecessor of the Fon used, the Fon still has to use because if a citizen of Mankon catches a leopard, tiger, buffalo and other wild animals, they carry it to the palace and when it is slaughtered they preserve its skin...and they remain there forever and can be destroyed only by fire and other similar disasters.... (Pa Tumanjong Paul)*

The most unusual of these are parts of animals, which fall under the category of protected wildlife. This is happening at a time when many people have hardly ever seen a live elephant (but on TV or in the zoo). These rare and exotic exhibitions are highly prized and do enable successive rulers to display their wealth and status.

2.16 The uses of wildlife/royal wildlife

The people believe that royalty owns royal wildlife. Royalty protects the land and needs to be informed of (honored) any wildlife killed; else, defaulters will suffer sanctions including at times excommunication from the village. Persons who distinguish themselves by capturing wildlife and taking it to the palace are decorated and given awards. One respondent stated that:

> *Of course when you have a red feather, given to you by the authority in the palace... you have some privileges which the common villager does not have because ... so that makes you different from the common man in the society...if you are a notable, let us say you have been given a red feather by the fondom...it is just like the chancellery having all the people who have had medal awards.... (Aborengong F., John; Food Market Mankon, November 2010)*

2.17 Wildlife and traditional medicine

Most people combine western medicine alongside traditional healing practices. They will, for example, consult western and traditionally trained medical practitioners simultaneously. They may also visit a medium (who will summon the spirits to treat the patient) even while staying in a hospital. Additionally, local remedies are known and used without consulting a healer.

Parts of plants and animals are, however, central features of traditional medicine. This is the reason for the demand of plant and animal materials for medicinal purposes. A significant proportion of this community use traditional medicine on a daily basis. The rarity of certain animal species has reduced the ability of the majority of the people of Mankon to afford to keep or purchase animal products for medicinal purposes. Vendors are increasingly selling fake animal products. Many traditional healers cannot afford to purchase these animal parts to include in their remedies. Some are substituting plant for animal ingredients or recommending to their clients that they source or purchase the animal-based ingredients themselves rather than relying on the healer. Some animal products are openly sold at traditional medicine outlets. These species are considered to be of a lower financial value, since the Forestry Law dictates a system of fines according to the perceived financial value of the species. Wildlife regulation has encouraged trading activities to continue "out of sight," and many wildlife products are now very expensive to buy from intermediaries. Traders prefer to buy them only when they have a ready customer (**Table 2**).

Some hunted wildlife species are highly valued for their medicinal properties. Examples are the elephant, tiger, and python, which are valued as therapy during labor and childbirth. Prior to the passing of the wildlife law, wildlife parts were readily available for sale in traditional medicine shops everywhere in the study region. Throughout the 1990s, Africa was widely considered to be a major source for tiger (*Panthera tigris*) parts for the international medicine markets [60–63]. Reports suggest that in the past the bones of tiger had only a limited domestic market and that the majority were shipped abroad for higher prices [60].

In the early 1990s, scarce elephant hooves were fetching USD10 per gram. When these same elephant parts became very scarce, they were being sold at USD700 [62]. Some informants however pointed out that the effects of market involvement on local wildlife populations and the humans who depend on them are varied and complex. The trade of wild animals generates income and employment, and it has the potential to help manage and regulate herd size and wildlife populations. These are useful by-products of wildlife marketing. On the other hand, poor management and overharvesting are two typical examples of how market forces exacerbated the trade in wildlife. Market involvement in the wildlife trade has created some problems, the opportunities, as well as incentives. The main problem is to find ways in which community forestry activities could maximize the positive aspects and minimize the negative impact of this involvement.

Animals	By-products used
Fowl	Blood, wings
Bee	Bee wax
Goat	Hoof
Cowries	Bone
White monkey	Spike
Porcupine	Bone
Leopard	Bone, hoof
Elephant	Bone
Black snake	Fat, bone
Python	Bone
Swallow	Bone
Fish	Bone
Owl	Bone
Duiker	Bone
Snail	Shell, flesh

Source: Fieldwork.

Table 2.
Wildlife and by-products used for medicinal purposes.

2.18 Wildlife as symbols in Mankon

Artists frequently use animals as a subject matter in their art. Animals are symbolic. Throughout the ages, many cultures have regarded specific animals as representing gods, power, and the supernatural. In Mankon, men (and especially the Fon) are believed to form alliances with animals in order to have a double, triple, or multiple existence and to have all the qualities, strengths, as well as weaknesses of the chosen animals to act in life efficiently. The Fon is often associated with the leopard, which is known for its ferocity, for its aggression, for being intrepid, as a great watcher, for being courageous, for being active, and for its speediness. He is also linked to the elephant, which is reputed for its strength, fidelity, memory, patience, wisdom, intelligence, and power [8]. As noted earlier, the Fon is mostly identified in public with the skin of a leopard and animal tusks which depicts this symbolic alliance with these animals projecting the qualities they are believed to incarnate:

> *All notables and clan heads use the horns of the buffalo as their drinking cups [...] it is a sign of superiority over the common villager [...] if you are not a notable you cannot be seen drinking from nothing more than a cow horn or a calabash cup [...] so the notables have the privilege of drinking with cups made from buffalos, the dwarf cows, and other strong animals as a sign of superiority among the villagers.... (PA Aboringong; food Market Mankon, November 2010)*

Mathias Alubafi [64] concedes that buffalo cow horn drinking cups decorated with Bruce Lee's facial image have become status objects and objects of decoration. Successful youths who have acquired these drinking horns use the fascination associated with it as "embodiments of a new iconography and iconology for the western Grassfields." By reversing "to their advantage the traditional iconography and iconology that was typical of the region and that was restricted to royals and elites" [64], they are questioning and reinterpreting cultural values.

Animals are regularly depicted on many packets, bottles, and sachets containing traditional medicine in the Mankon marketplace. According to medicine men, animal images and names do not indicate the ingredients in the medicine. They are, however, used as an attraction and for branding to indicate the effectiveness of the medicine. They belief that medicines containing animal derivatives are effective appear to be deeply entrenched in the local community. Similarly, a survey carried out in 1994 found that the popularity of a so-called Loris wine dropped when it was rumored to be mass produced without the use of Loris ingredients [65]. A medicine man in the heart of Mankon stated that he had given up using animal products on grounds that they were too expensive and too hard to find. Indeed, with the exception of one tonic containing an unspecified toad, no evidence of the use of animals in his retail premises or workshop was found. However, on the label of his plant based post-partum tonic could be read: "Works in the same way as porcupine stomach." Another medicine man, who has shops in Mankon food market and sells in other periodic markets like in Momo and Bui Divisions, also uses animals as a marketing strategy. He stated that in advertising his products, he uses slogans like, "No synthetic substances are used in my remedies, only natural ingredients, such as plants and the skin and gallbladders of animals." Respondents stated that the wildlife of the community is considered the people's cultural heritage. However, through overhunting, timber harvesting, bush fires, the use of toxic chemicals and other forms of habitat destruction, a large number of animals, reptiles, and most bird species have become extinct (see **Table 3**).

2.19 Relationship between the Mankon community and wildlife

While animals do not have an exact incredible large influence on other species and the natural world like humans do, there is increasing fear of the extinction of innumerable wildlife species. If the present trend of global human population is large enough and the technologies that allow humans to manipulate the environment are potent enough, human-caused alterations to the biosphere could fasten extinction. Mankon is far from hunter-gatherer societies that obtain their food directly from natural ecosystems, by hunting wild animals and collecting wild plants. It is noteworthy that a hunter-gatherer lifestyle can support a relatively can support a relatively small number of people in most landscapes. So population densities of such societies tend to be low. Contrary to hunter-gatherer societies, Mankon has long mutated from an exclusive farming community to an agrarian society, which obtains food not just by foraging in natural ecosystems but also by planting species that are important food items and/or raising livestock. Contrary to most hunter-gatherer societies where people are migratory, traveling frequently in search of food rather than living in settlements, the Mankonians supplement the food they raise with hunting and foraging adapted to the species still prominent in their environment. Mankon has gradually evolved from a strictly agrarian and pastoralist society to incarnate the two forms of farming [65].

The poor soil of this region usually cannot support the permanent, large-scale, plowed farming style of more advanced agricultural societies. Hence, they practice small-scale slash and burn agriculture meant for family consumption purposefully with the rest sold for the purchase of other basic needs. However, in the olden days, animals were commonly associated with particular gods and goddesses and were often symbolic of a deity's power. Encountering a particular species of wildlife may be considered as an omen from a god, but the power usually did not reside in the animal itself but rather in its relationship to a deity. However, horticultural and herding societies are generally confined to only certain climates and habitat types, and their population densities are still relatively low; so often with some societies,

Animal name		Mankon name	Type
1	Leopard	*afon*	Carnivores
2	Lion	*sa-mbang*	Carnivores
3	Hyena	*angwaba'/takfura*	Carnivores
4	Elephant	*Ǘsen*	Herbivores
5	Waterbuck	*anjwah*	Herbivores
6	Situntunga	*angyie*	Herbivores
7	Antelope	*angab*	Herbivores
8	Hare	*chwi*	Herbivores
9	Civit cat	*atsob*	Herbivores
10	Porcupine	*njibnǘ*	Herbivores
11	Monkey	*nkan*	Herbivores
13	Wild cat	*mǘmang*	Herbivores
14	Otter	*nyamnkyi*	Herbivores

Source: Field notes.

Table 3.
List of identified rare or extinct in Mankon.

there are still considerable undisturbed areas that provide habitat for wildlife. This is the case with Mankon where vast extensive land areas are yet to be exploited for farming. The transition from hunter-gatherer to early agrarian societies often resulted in increased birth rates. Growing human populations place ever-increasing demands on the surrounding wildlife and natural ecosystems.

Furthermore, there are reports of the increasing scarcity for many medicinal wildlife species. This situation represents a concern not only from the conservation point of view but also because reduced availability of medicinal wildlife will have a negative effect on the health status of many people living in this community in particular and in Cameroon as a whole. There is certainly a threat to the medicinal importance of wildlife in that scarce wildlife population will not be able to ensure continuous supply of wildlife parts/products for medicines. Concurrently, there will not be a continuous use of wildlife in traditional drug development and the supply of wildlife for spiritual and ceremonial purposes.

2.20 Wildlife use in Mankon and hunting methods

They include legal and illegal as well as sustainable and unsustainable means. Three primary modes of wildlife hunting are currently taking place in Mankon and resulting in three different products—bushmeat, game meat, and wild meat. Apart from the bushmeat trade, there are other forms of wildlife harvest that contribute to meeting the nutritional, economic, and other cultural needs of the Mankon community in particular and African communities in general. The wildlife trade across the Northwest Region and Mankon in particular has rapidly evolved over recent years to become a lucrative and culturally oriented phenomenon. Its dramatic expansion has been facilitated by numerous factors including increasing consumer demand, development of road networks and industrial infrastructure (logging, mining, oil), market declines in other sectors (e.g., agricultural), increasing human population, lack of incentive or resources to develop alternatives, social and political upheavals, inadequate monitoring and enforcement capacity, and decreases in household economic security. Reports indicate that there is a significant, illegal, commercial exploitation of wildlife occurring around the globe and that current levels of exploitation are unsustainable and threaten the future of numerous wildlife populations and the people dependent on them [66–70]. Game meat which describes legally obtained meat as part of a commercial (private or communally managed) operation that is regulated and controlled is also where monitoring of the wildlife populations and habitat is carried out and where trade is legally conducted with authorized agents and government controls [66–69]. The game meat trade results from a form of uniquely managed wildlife exploitation called game ranching involving both consumptive and nonconsumptive activities that are more developed in selected countries in East and Southern Africa. Its development was formally established in the 1960s in Southern Africa and has grown in recent years as a recommended alternative land-use strategy that provides increased ecological and economic benefits to cattle ranching or crop farming. Populations of wildlife are maintained within (un)fenced areas and often involve the interchange of genetic stock with unmanaged populations (i.e., wild). Production systems may be privately or communally managed and involve species of birds and mammals. The presence of predators is allowed in some ranching systems but not in all [65–67]. Though not well organized, this kind of game meat management is highly perceptible in Mankon. Game cropping is one consumptive use performed on game ranches for the purposes of producing meat, skins, and horns for sale in either domestic or international markets. It requires considerable management in both biological and economic terms for the maximization of efficiency [43, 67–70]. Elderly respondents informed us that small-scale game cropping had existed in the Mankon area. Additional activities found on game ranches may include sport hunting, wildlife viewing, and scientific research. The game ranching and game meat industry outside of savannah ecosystems are not being widely developed. Within these regions, the advantages and disadvantages of such efforts raise doubts as to the ecological, economic, and social viability of such operations. Many observers believe that game ranching operations are sustainable, but they require a significant management commitment and usually produce limited economic returns with only a few select species being viable for limited commercial exploitation. Wild meat in Mankon is for subsistent noncommercial purposes. For purposes of legal hunting, traditional snares, nets, registered guns, traps, etc. are used. It is undertaken by individuals with the legal rights to access the wildlife for purposes of subsistent and legal sport hunting. Wild meat includes only those species legally authorized for harvest and may be used for local trade of basic needs items—clothes, carbohydrates, and household products [43].

2.21 Gender and wildlife in Mankon

Gender permeates and is an important consideration in all development and conservation work. While a lot is known about the nature and extent of gender roles with respect to agricultural and forestry planning, participation, and decision-making, much less is known about the gender roles of local peoples with respect to the management of wild animal species. The traditional gender division of labor identifies women as child bearers, responsible for nurturing and healing within household and in the community as well. In order to fulfill this socially defined function, women have a special dependence, which often differs from that of men, on the natural resources around them that they use and often manage [70]. Thus beyond their accepted traditional roles, women are also often the invisible managers and decision-makers within both the private sphere of the household and the public sphere of the community. To ensure that both women and men are active participants in and beneficiaries of sustainable wildlife use, it is imperative to consider them as equal partners from the outset. This becomes especially important when considering the incorporation of wild animal species into specific projects or activities. For example, in light of the food taboos identified in the community, one needs to ask whether the animals incorporated in such a project are also beneficial to women or not. Are the animals useful resources for men and women? Will they threaten the home gardens for which women have responsibility? Should areas containing identified extinct wildlife species be earmarked for community forests? As Hoskins [70] and others suggest, it is important to consider the role of women and men in planning, participating, and benefiting from forestry activities. Hoskins illustrates this point with an example of beekeepers in Kenya. It was found that a beekeeping project in that country was receiving no support from women, until the project director realized that it was culturally unacceptable for women to climb trees, and so they were not able to reach the beehives. Once beehives were placed close to the ground, women became willing participants.

Women and men in rural societies like Mankon interact with wildlife in other important ways. Observations showed that in Mankon women are gatherers of forest products (usually non-animal products or small animals such as insects). Men are the hunters of the larger wild animal species. Although men are the primary hunters of the large game, women are frequently involved in catching, butchering, and transportation of animals as well as in cooking and preservation of their meat. In Mankon, for example, it is usually women who snare or trap small animals, such as rabbits, for food and fur; it is they who clean, dry, and smoke meat and fish; and it is they who scrape, clean, and tan hides for clothing and rawhide. This contrasts to the Zambian experience where, although women rarely joined men in actual elephant hunts, they perform many vital activities as part of the overall elephant exploitation process.

3. Conclusions

This paper has explored the perceptions of the Mankon community members on wildlife use and its entanglement with their sociocultural practices and how the lack of cultural sensitivity in the formulation of Cameroon's 1994 Forestry Law exposes the people to legal sanctions and threatens their cultural well-being. The paper has demonstrated that parts of particular wildlife species are used for medicinal purposes and as symbols of authority because of their cultural association with certain intrinsic values. Certain behaviors including taboos that are linked with the hunting of particular wildlife species have implications for the culture of this community. This explains the community's myriad efforts at conservation. Other local mechanisms of conservation include totemism, and the meticulous timing of when, the quantity and the individual entitled to harvest particular plant species.

There is a need for legal change to guarantee the enjoyment of indigenous land rights including:

1. By moving Cameroonian legislations beyond its focus on the farm. This can be achieved through the granting of collective land rights for marshlands, pastures, rangelands, forests, and woodlands and not just the recognition of farms (*mis en valeur principle*) [26–29].

2. Recognizing customary land rights as legal rights of ownership and granting the same level of protection as for lands held under introduced nonindigenous systems [26, 33].

A more inclusive and effective forestry policy that protects species should link the conservation of culture and nature rather than a strategy that ignores traditional institutions, knowledge systems, and practices. Taking into consideration local cultural traditions are in line with enhancing Cameroon's compliance with various biodiversity-related international environmental conventions that the country has endorsed. We recommend the incorporation of sacred forests and shrines into the protected area system and the granting of the rights of local peoples in these spaces.

There is also the need for the education of the masses about sustainable environmental management practices. It has been demonstrated that an increase in people's education and therefore their awareness level is correlated with higher levels of environmental protection and that educated individuals "are more likely to generate an environmentally progressive civil service, and therefore have democratically minded public policymakers and organisations that are more receptive to public demands for environmental protection" ([70], p. 8). On the other hand, the lack of education is perceived as an obstacle to "public understanding and awareness of environmental issues" [4, 14, 71, 72].

Environmental education as succinctly put by the World Conservation Union encompasses the elements of behavior and the idea that knowledge will induce personal, societal, and global changes [11–13]. It will afford learners an "opportunity to gain an awareness or sensitivity to the environment, knowledge and experience of the problems surrounding the environment, to acquire a set of values and positive attitudes, to obtain the skills required to identify and solve environmental problems and, the motivation and ability to participate" ([14], p. 9). A gender lens will enhance women's participation in the management of natural resources and make them cooperate in sustainable development that will ensure the conservation of scarce fauna and flora species.

Acknowledgements

We sincerely and wholeheartedly thank various respondents who volunteered their time and taught us the culture of the Mankon people of the Western Grassfield Region of Cameroon during various ethnographic trips to the region. We appreciate their hospitality.

Conflict of interest

The authors have no conflict of interest.

Author details

Ngambouk Vitalis Pemunta[1*] and Ngwa Donald Anye[2]

1 Department of Public and Community Health Medicine, Institute of Medicine, Sahlgrenska Academy of the University of Gothenburg, Gothenburg, Sweden

2 Department of Anthropology, University of Yaounde 1, Yaounde, Cameroon

*Address all correspondence to: vitalis.pemunta@gu.se

References

[1] Ripple WJ, Abernethy K, Betts MG, Chapron G, et al. Bushmeat hunting and extinction risk to the world's mammals. Royal Society Open Science. 2016;**3**(10):160498

[2] Butynski TM. African Primates: The Newsletter of the Africa Section of IUCN/SSC Primate Specialist Group. Vol. 3. 1997. pp. S1-S42

[3] Ministère des forêts et de la faune (MINFOF). Direction de la faune et des aires protégées. L'Arrêté no 0648/MINFOF du 18 décembre 2006 fixant la liste des animaux des classes de protection A, B, C. Nous y ferons référence en tant que "l'Arrêté". 2013. Available from: http://laga-enforcement.org/Portals/0/Documents/Legal%20documents/Cameroon/l%C3%A9gislation%20faunique_Protection_esp%C3%A8ce_menac%C3%A9es-Fr_En.pdf [Accessed: June 20, 2018]

[4] International Union for Conservation of Nature and Natural Resources (IUCN). Red List of Threatened Species Version. 2010. Available from: www.iucnredlist.org

[5] Ingram V, Jam NA. Plant and Animal Guide for the Western Cameroon Highlands. Technical Report; SNV. 2007

[6] Anye ND. Abŭng a Fo: The mankon annual dance festival, Maitrise [thesis]. Cameroon: Department of Anthropology, University of Yaounde 1; 2004

[7] Usongo LI. Conservation status of primates in Cameroon. Primate Conservation. 1998;**18**:59-65

[8] Notue JP. The Treasure of the Mankon Palace: Cultural Objects from the Royal Palace. Balmayo: Institut de Formation Artistique; 2000

[9] Nelson RR, Edmund SP. Investment in Humans, technological diffusion, and economic growth. The American Economic Review. 1966;**56**(1 and 2): 69-75

[10] Oxford English Dictionary. The Oxford English Paperback Dictionary. Oxford, UK: Oxford University Press; 1990

[11] Howe C. The Role of Education as a Tool for Environmental Conservation and Sustainable Development, A [dissertation submitted for the degree of Doctor of Philosophy]. Imperial College London; 2009

[12] Vin-Mbah FI. Learning and teaching methodology. Journal of Educational and Social Research. 2012;**2**(4):111-118

[13] International Union for the Conservation of Nature (IUCN). International Working Meeting on Environmental Education in the School Curriculum, Final Report. Tech. Rep. IUCN; 1970

[14] Jacobson S, McDuff M, Monroe M. Conservation Education and Outreach Techniques. Oxford, UK: Oxford Biology; 2006

[15] Marcus GE. Ethnography in/of the world system: The emergence of multi-sited ethnography. Annual Review of Anthropology. 1995;**24**:95-117

[16] Marcus GE. Ethnography Through Thick and Thin. Princeton, NJ: Princeton University Press; 1998

[17] Curtis E, Drennan J. Quantitative Health Research: Issues and Methods. Maidenhead: Open University Press; 2013

[18] Probyn JE, Howarth ML, Maz J. The 'middle bit': How to appraise qualitative, research. British Journal of

Cardiac Nursing. 2016;**11**(5):1-20. DOI: 10.12968/bjca.2016.11.5.248

[19] Palinkas LA, Horwitz SM, Green C, Wisom JP, Duan N, Hoagwood K. Purposeful sampling for qualitative data collection and analysis in mixed method implementation research. Administration and Policy in Mental Health. 2015;**42**(5):533-544

[20] Polit DF, Beck CT. Essentials of Nursing Research: Appraising Evidence for Nursing Practice. 8th ed. Philadelphia: Wolters Kluwer; 2014

[21] Austin ZA, Sutton J. Qualitative research: Getting started. The Canadian Journal of Hospital Pharmacy. 2014;**67**(6):436-440

[22] Boydell KM, Volpe T, Pignatiell A. A qualitative study of young people's perspectives on receiving psychiatric services via Televideo. Journal of Canadian Academy of Child and Adolescent Psychiatry. 2010;**19**(1):5-11

[23] Bernard RH. Research Methods in Anthropology: Qualitative and Quantitative Approaches. 4th ed. New York: Altamira Press; 2007

[24] Nguiffo S, Talla M. Cameroon's wildlife legislation: Local custom versus legal conception. Unasylva. 2010;**236**:61

[25] Mamdani M. Citizen and Subject: Contemporary Africa and the Legacy of Late Colonialism (with Preface by the Author). Princeton: Princeton University Press; 2018

[26] Warnier J-P. La Mondialisation de la Culture. Paris: Edition la Decourverte et Syros; 1999

[27] Njoh AJ. Continuity and change in Cameroonian land policy. Planning Perspectives. 2000;**15**(3):241-265

[28] Hoagland SJ. Integrating traditional ecological knowledge with Western science for. Optimal Natural Resource Management. 2017;**3**(1):1-15. DOI: 10.18113/P8ik359744

[29] Yerima BPK. Prototype for Watershed Management and Biodiversity Conservation in the Western Highland Regions of Cameroon; Victoria BC, Canada. 2012. p. 440

[30] Pemunta NV, Mbu-Arrey OP. The tragedy of the governmentality of nature: The case of national parks in Cameroon. In: Bigman D, Smith JB, editors. National Parks: Sustainable Development, Conservation Strategies, and Environmental Effects. Georgia, United States: Nova Science Publishers; 2013. pp. 1-46

[31] Fru BS. Contribution of Traditional Institutions on the Sustainable Management of Sacred Forests: Case Study of Mankon Sacred Forests, Northwest Region, Cameroon, Master of Science (M.Sc.) [thesis]. Environmental Management, University of Dschang; 2016

[32] Oyono PR. The foundations of the conflit de langage over land and forests in southern Cameroon. African Study Monographs. 2005;**26**(3):115-144

[33] Wily L. Whose Land Is it? The Status of Customary Land Tenure in Cameroon. New York: Redlin Print Limited; 2011

[34] Pemunta NV, Njiki M. Experiencing neoliberalism from below: The Bakweri confrontation of the state of Cameroon over the privatization of the Cameroon development corporation. Journal of Human Security. 2010;**6**(1):38-53

[35] Pemunta NV. Fortress conservation, wildlife legislation and the Baka pygmies of Southeast Cameroon. GeoJournal. Published online First; 2018. DOI: 10.1007/s10708-018-9906-z

[36] Sama SM. Essential Readings in Environmental Law IUCN Academy of Environmental Law (www.iucnael.org). Environmental Law in Cameroon, 2014. Available from: www.Environmental%20Law%20in%20Cameroon.Dec.5.2014.final.pdf [Accessed: January 20, 2018]

[37] Vaccaro I, Paquet P. Political ecology and conservation policies: Some theoretical genealogies. Journal of Political Ecology. 2013;**20**(20):255-272

[38] Peet R, Watts M. Liberation ecology: Development, sustainability, and the environment in an age of market triumphalism. In: Peet R, Watts M, editors. Liberation ecologies: Environment, development, social movements. London: Routledge; 2004. pp. 1-47

[39] Smith AE, Wishnie M. Conservation and subsistence in small-scale societies. Annual Review of Anthropology. 2000;**21**(1):493-524

[40] Egute TO. Modern Law and Local Tradition in Forest Heritage Conservation in Cameroon: The Case of Korup Ph.D [Dissertation]. Senftenberg, Germany: Environmental Sciences, Brandenburg University of Technology Cottbus; 2102

[41] Alcorn JB. Indigenous peoples and conservation. Conservation Biology. 1993;7(2):424-426

[42] Alvard MS. Evolutionary ecology and resource conservation. Evolutionary Anthropology. 1998;7(2):62-74

[43] Eves HE, Ruggiero RG. Socio-economics and the sustainability of bushmeat hunting in the forests of Northern Congo (Brazzaville). In: Robinson JG, Bennett EL, editors. Hunting for Sustainability in Tropical Forests. New York: Columbia University Press; 2000. pp. 427-454

[44] Eballa Y, Angamo AE. Some Notes on Zintgraff's Punitive Expedition against the Mankon People (Excerpts, Translations and Comments). Focus on Niikwi Nii Fo NdeFru III. Mankon Cultural Festival; 23rd–31st December 1884. 2013. pp. 34-45

[45] Nkwi PN, Warnier J-P. Elements of a History of the Western Grassfields. Yaounde: ICARSST; 1982

[46] Mah A-a V. Sustainability of community-managed projects in the North West Region of Cameroon [thesis]. Cardiff Metropolitan University; 2016

[47] Mbiti J. African Religions and Philosophy. London: Heinemann; 1990

[48] Byers BA, Cunliffe R, Hudak AT. Linking the conservation of culture and nature: A case study of sacred forests in Zimbabwe. Journal of Human Ecology. 2001;**29**:187-218

[49] Douglas M. Purity and Danger: An Analysis of the Concepts of Pollution and Taboo. London and New York: Routledge Classics; 1966

[50] Redford KH, Godshalk R, Asher K. What About the Wild Animals? Wild Animal Species in Community Forestry in the Tropics. FAO Corporate Document Repository; 1995

[51] Ntaimah TP. The future of medicinal plants in a nearly extinct kilum mountain forest reserve in Oku-Bui. An inside exploration into the cultural Beliefs and Practices of the Oku people on conservation of Medicinal plants, Masters in Anthropology [thesis]. Cameroon: University of Yaounde 1; 1997

[52] Faure V. Bodies and Politics: Healing Rituals in the Democratic South Africa. Les Cahiers de L'IFAS No. 2. French Institute of South Africa; 2002

[53] Lewellen TC. Political Anthropology, An Introduction. Westport CT: Bergin & Garvey; 1992

[54] Ketzer DI. Politics and Symbols: The Italian Communist Party and the Fall of Communism. New Haven and London: Yale University Press; 1996

[55] Bailey RC, Bahuchet S, Hewlett B. Development in the central African Forest: Concern for peoples. In: Cleaver K, Munasinghe M, Dyson N, Egli A, Peuker A, Wencelius M, editors. Conservation of West and Central African Rainforests. World Bank Environment Paper No. 1. Washington, DC: The World Bank; 1992. pp. 202-211

[56] Assitou, Sidle. Congo game conservation: Limits of a legal approach. In: Bissonette JA, Krausman PR, editors. Integrating People and Wildlife for a Sustainable Future. Bethesda, MD, USA: The Wildlife Society; 1995. pp. 509-511

[57] Carpaneto GM, Germi F. Diversity of mammals and traditional hunting in central African rain forests. Agriculture, Ecosystems and Environment. 1992;**40**:335-354

[58] Kitanishi K. The impact of cash and commoditization on the Baka hunter-gatherer society in southeastern Cameroon. African Study Monographs. 2006;**33**:121-142

[59] Southern Africa Sustainable Use Specialist Group (SASUSG), The IUCN Regional Office for Southern Africa (IUCN ROSA), The International Institute for Environment and Development (IIED). Community Wildlife Management in Southern Africa. A Regional Review. Evaluating Eden Series Paper No. 11; 1997

[60] Cooney RD, Dilys H, Holly W, David K, Aidan T, Henry S, et al. From poachers to protectors: Engaging local communities in solutions to illegal wildlife trade. Policy Perspectives. 2016;10(3):367-374. DOI: 10.1111/conl.12294

[61] Naughton-Treves L, Barndon K, Holland M. Role of protected areas in conserving biodiversity and sustaining local livelihoods. Annual Review of Environment and Resources. 2005;**17**(30):219-252

[62] Nowell K. Far from a Cure: The Tiger Trade Revisited. Cambridge, UK: Traffic International; 2000

[63] Martin EB, Phipps M. A review of the wild animal trade in Cambodia. TRAFFIC Bulletin. 1996;**16**(2):45-60

[64] Alubafi MF, Rampahlile M, Valkenburg G. The shifting iconography of drinking horns in the Western Grassfields, Cameroon. Cogent Social Sciences. 2017;**3**(1):3, 1375598. DOI: 10.1080/23311886.2017.1375598

[65] Richerson PJ, Boyd R, Bettinger RL. Was agriculture impossible during the pleistocene but mandatory during the Holocene? A climate change hypothesis. American Antiquity. 2001;**66**(3):387-411

[66] Bennett EL, Robinson JG. Hunting of wildlife in tropical forests: Implications for biodiversity and forest people WCS New York. In: Bennett EL, Robinson JG, editors. Hunting for Sustainability in Tropical Forests. New York: Columbia University; 2000

[67] Geist V. How markets in wildlife meat and parts, and the sale of hunting privileges, jeopardize wildlife conservation. Conservation Biology. 1988;**2**(1):15-26. DOI: 10.1111/j.1523-1739.1988.tb00331.x

[68] Swanson TM, Barbier E, editors. Economics for the Wilds: Wildlife, Diversity, and Development. Washington, D.C.: Island Press; 1992

[69] Sommerlatte M, Hopcraft D. The economics of game cropping on a Kenyan ranch. In: Proceedings of the 3rd International Wildlife Ranching Symposium; Pretoria, Republic of South Africa. 1992

[70] Hoskins M. The contributions of forestry to food security. Unasylva. 1990;**41**:3-13

[71] Kinda SR. Essays on Environmental Degradation and Economic Development. Economies and Finances. Université d'Auvergne-Clermont-Ferrand I. English. 2013

[72] Mitra A. Rising population and environmental degradation. Yojana. 1984;**28**(18):4-8